T0336037

## Applied Speech and Audio Processing: With MATLAB® Examples

*Applied Speech and Audio Processing* is a MATLAB-based, one-stop resource that blends speech and hearing research in describing the key techniques of speech and audio processing.

This practically orientated text provides MATLAB examples throughout to illustrate the concepts discussed and to give the reader hands-on experience with important techniques. Chapters on basic audio processing and the characteristics of speech and hearing lay the foundations of speech signal processing, which are built upon in subsequent sections explaining audio handling, coding, compression and analysis techniques. The final chapter explores a number of advanced topics that use these techniques, including psychoacoustic modelling, a subject which underpins MP3 and related audio formats.

With its hands-on nature and numerous MATLAB examples, this book is ideal for graduate students and practitioners working with speech or audio systems.

**Ian McLoughlin** is an Associate Professor in the School of Computer Engineering, Nanyang Technological University, Singapore. Over the past 20 years he has worked for industry, government and academia across three continents. His publications and patents cover speech processing for intelligibility, compression, detection and interpretation, hearing models for intelligibility in English and Mandarin Chinese, and psychoacoustic methods for audio steganography.

# Applied Speech and Audio Processing

## With MATLAB® Examples

IAN MCLOUGHLIN

School of Computer Engineering
Nanyang Technological University
Singapore

# CAMBRIDGE
## UNIVERSITY PRESS

University Printing House, Cambridge CB2 8BS, United Kingdom

Cambridge University Press is part of the University of Cambridge.

It furthers the University's mission by disseminating knowledge in the pursuit of education, learning and research at the highest international levels of excellence.

www.cambridge.org
Information on this title: www.cambridge.org/9780521519540

© Cambridge University Press 2009

First published 2009

*A catalogue record for this publication is available from the British Library*

ISBN 978-0-521-51954-0 Hardback

# Contents

# Preface

Speech and hearing are closely linked human abilities. It could be said that human speech is optimised toward the frequency ranges that we hear best, or perhaps our hearing is optimised around the frequencies used for speaking. However whichever way we present the argument, it should be clear to an engineer working with speech transmission and processing systems that aspects of both speech and hearing must often be considered together in the field of vocal communications. However, both hearing and speech remain complex subjects in their own right. Hearing particularly so.

In recent years it has become popular to discuss psychoacoustics in textbooks on both hearing and speech. Psychoacoustics is a term that links the words *psycho* and *acoustics* together, and although it sounds like a description of an auditory-challenged serial killer, actually describes the way the mind processes sound. In particular, it is used to highlight the fact that humans do not always perceive sound in the straightforward ways that knowledge of the physical characteristics of the sound would suggest.

There was a time when use of this word at a conference would boast of advanced knowledge, and familiarity with cutting-edge terminology, especially when it could roll off the tongue naturally. I would imagine speakers, on the night before their keynote address, standing before the mirror in their hotel rooms practising saying the word fluently. However these days it is used far too commonly, to describe any aspect of hearing that is processed nonlinearly by the brain. It was a great temptation to use the word in the title of this book.

The human speech process, while more clearly understood than the hearing process, maintains its own subtleties and difficulties, not least through the profusion of human languages, voices, inflexions, accents and speaking patterns. Speech is an imperfect auditory communication system linking the meaning wishing to be expressed in one brain, to the meaning being imparted in another brain. In the speaker's brain, the meaning is encoded into a collection of phonemes which are articulated through movements of several hundred separate muscles spread from the diaphragm, through to the lips. These produce sounds which travel through free air, may be encoded by something such as a telephone system, transmitted via a satellite in space half way around the world, and then recreated in a different environment to travel through free air again to the outer ears of a listener. Sounds couple through the outer ear, middle ear, inner ear and finally enter the brain, on either side of the head. A mixture of lower and higher brain functions then, hopefully, recreate a meaning.

It is little wonder, given the journey of meaning from one brain to another via mechanisms of speech and hearing, that we call for both processes to be considered together. Thus, this book spans both speech and hearing, primarily in the context of the engineering of speech communications systems. However, in recognition of the dynamic research being undertaken in these fields, other areas are also drawn into our discussions: music, perception of non-speech signals, auditory scene analysis, some unusual hearing effects and even analysis of birdsong are described.

It is sincerely hoped that through the discussions, and the examples, the reader will learn to enjoy the analysis and processing of speech and other sounds, and appreciate the joy of discovering the complexities of the human hearing system.

In orientation, this book is unashamedly practical. It does not labour long over complex proofs, nor over tedious background theory, which can readily be obtained elsewhere. It does, wherever possible, provide practical and working examples using MATLAB to illustrate its points. This aims to encourage a culture of experimentation and practical enquiry in the reader, and to build an enthusiasm for exploration and discovery. Readers wishing to delve deeper into any of the techniques described will find references to scientific papers provided in the text, and a bibliography for further reading following each chapter.

Although few good textbooks currently cover both speech and hearing, there are several examples which should be mentioned at this point, along with several narrower texts. Firstly, the excellent books by Brian Moore of Cambridge University, covering the psychology of hearing, are both interesting and informative to anyone who is interested in the human auditory system. Several texts by Eberhard Zwicker and Karl D. Kryter are also excellent references, mainly related to hearing, although Zwicker does foray occasionally into the world of speech. For a signal processing focus, the extensive Gold and Morgan text, covering almost every aspect of speech and hearing, is a good reference.

## Overview of the book

In this book I attempt to cover both speech and hearing to a depth required by a fresh postgraduate student, or an industrial developer, embarking on speech or hearing research. A basic background of digital signal processing is assumed: for example knowledge of the Fourier transform and some exposure to discrete digital filtering. This is not a signal processing text – it is a book that unveils aspects of the arcane world of speech and audio processing, and does so with MATLAB examples where possible. In the process, some of the more useful techniques in the toolkit of the audio and speech engineer will be presented.

The motivation for writing this book derives from the generations of students that I have trained in these fields, almost each of whom required me to cover these same steps in much the same order, year after year. Typical undergraduate courses in electronic and/or computer engineering, although they adequately provide the necessary foundational skills, generally fail to prepare graduates for work in the speech and audio signal processing field. The coverage in this book is targeted toward filling the gap. It is designed to educate, interest and motivate researchers working in this field to build their skills and capabilities to prepare for research and development in the speech and audio fields.

This book contains seven chapters that generally delve into deeper and more advanced topics as the book progresses. Chapter 2 is an introductory background to basic audio processing and handling in MATLAB, and is recommended to those new to using MATLAB for audio work. It also contains justifications for, and explanations of, segmentation, overlap and windowing, which are fundamental techniques in splitting up and handling long recordings of speech and audio.

Chapter 3 describes speech production, characteristics, understanding and handling, followed by Chapter 4 which repeats the same for hearing. Chapter 5 is concerned with the handling of audio, primarily speech, and Chapter 6 with analysis methods for speech and audio. Finally Chapter 7 presents some advanced topics that make use of many of the techniques in earlier chapters.

## Arrangement of the book

Each section begins with introductory text explaining the points to be made in the section, before further detail, and usually MATLAB examples are presented and explained. Where appropriate, numbered citations will be provided to a reference list at the end of each chapter. A bibliography is also provided at the end of each chapter, containing a set of the most useful texts and resources to cover the major topics discussed in the text.

---

**Infobox 0.1** Further information

---

Self-contained items of further interest, but not within the flow of the main text, are usually placed inside an infobox like this one for rapid accessibility.

---

Commands for MATLAB or computer entry are written in a typewriter font to distinguish them from regular text:

```
type this in MATLAB
```

All of the MATLAB commands are designed to be typed into the command window, or included as part of an m-file program. This book will not use Simulink for any of the examples, and will attempt to limit all examples to the basic MATLAB without optional toolboxes wherever possible.

It is my sincere hope that academics and industrial engineers alike will benefit from the practical and hands-on MATLAB approach taken in this book.

MATLAB® is the registered trademark of MathWorks, Inc. All references to MATLAB throughout this work should be taken as referring to MATLAB®.

# Acknowledgements

Kwai Yoke, Wesley and Vanessa graciously gave up portions of their time with me whilst I worked on this text. My parents encouraged me, not just for the writing (it's not as easy as it may appear), but also for my career in research and my education in general.

For my initial interest in speech and hearing, I must thank many friends and role models from HMGCC, the University of Birmingham and Simoco Telecommunications in Cambridge. In particular, Dr H. Ghafouri-Shiraz who guided me, helped me, encouraged and most of all, led by example. His own books on laser diodes and optical fibres are essential reading in those fields, his research skills are exceptional and his teaching exemplary. I would also like to thank Jim Chance for his guidance, help and supervision during my own PhD studies.

More recently, sincere thanks are due to Doug McConnell of Tait Electronics Ltd, Christchurch, and management guru Adrian Busch, for more than I could adequately explain here. The multitalented Tom Scott and enthusiastic Stefan Lendnal both enriched my first half decade in New Zealand, and from their influence I left, hopefully as a better person.

Hamid Reza Sharifzadeh kindly proofread this manuscript, and he along with my other students, constantly refined my knowledge and tested my understanding in speech and audio. In particular I would like to acknowledge the hard work of just a few of my present and past students in this field: Farzane Ahmadi, Fong Loong Chong, Ding Zhongqiang, Fang Hui, Robertus Wahendro Adi and Cedric Tio.

Moving away from home, sincere thanks are due to the coffee growers of the world who supported my writing efforts daily through the fruits (literally) of their labours.

Above all, everything in me that I count as good comes from the God who made me and leads me: all honour and glory be to Him.

# 1 Introduction

Audio and speech processing systems have steadily risen in importance in the everyday lives of most people in developed countries. From 'Hi-Fi' music systems, through radio to portable music players, audio processing is firmly entrenched in providing entertainment to consumers. Digital audio techniques in particular have now achieved a domination in audio delivery, with CD players, Internet radio, MP3 players and iPods being the systems of choice in many cases. Even within television and film studios, and in mixing desks for 'live' events, digital processing now predominates. Music and sound effects are even becoming more prominent within computer games.

Speech processing has equally seen an upward worldwide trend, with the rise of cellular communications, particularly the European GSM (Global System for Mobile communications) standard. GSM is now virtually ubiquitous worldwide, and has seen tremendous adoption even in the world's poorest regions.

Of course, speech has been conveyed digitally over long distance, especially satellite communications links, for many years, but even the legacy telephone network (named POTS for 'Plain Old Telephone Services') is now succumbing to digitisation in many countries. The *last mile*, the several hundred metres of twisted pair copper wire running to a customer's home, was never designed or deployed with digital technology in mind, and has resisted many attempts over the years to be replaced with optical fibre, Ethernet or wireless links. However with DSL (digital subscriber line – normally asymmetric so it is faster in one direction than the other, hence ADSL), even this analogue twisted pair will convey reasonably high-speed digital signals. ADSL is fast enough to have allowed the rapid growth of Internet telephony services such as Skype which, of course, convey digitised speech.

## 1.1 Digital audio

Digital processing is now the method of choice for handling audio and speech: new audio applications and systems are predominantly digital in nature. This revolution from analogue to digital has mostly occurred over the past decade, and yet has been a quiet, almost unremarked upon, change.

It would seem that those wishing to become involved in speech, audio and hearing related research or development can perform much, if not all, of their work in the digital domain these days. One of the benefits of digital technology is that the techniques are relatively device independent: one can create and prototype using one digital processing platform, and then deploy upon another platform. The criteria then for a development platform would be for ease-of-use and testing, while the criteria for a deployment platform may be totally separate: low power, small size, high speed, low cost, etc.

In terms of development ease-of-use, MATLAB running on a PC is chosen by many of those working in the field. It is well designed to handle digital signals, especially the long strings of audio samples. Built-in functions allow most common manipulations to be performed easily, audio recording and playback are equally possible, and the visualisation and plotting tools are excellent. A reduced-price student version is available which is sufficient for much audio work. The author runs MATLAB on both Mac OS-X and Linux platforms for much of his own audio work.

Although there is currently no speech, audio or hearing toolbox provided by The MathWorks® for MATLAB, the Signal Processing Toolbox contains most of the required additional functions, and an open source VOICEBOX is also available from the Department of Electrical and Electronic Engineering, Imperial College, London with many additional useful functions. It is also possible to perform all of the audio and speech processing in this book using the open source developed Octave environment, but would require some small changes to the MATLAB examples given. In terms of capabilities, Octave is less common than MATLAB, lacks the advanced plotting and debugging capabilities, but is otherwise similar.

## 1.2    Capturing and converting sound

This book is all about sound. Either sound created through the speech production mechanism, or sound as heard by a machine or human. In purely physical terms, sound is a longitudinal wave which travels through air (or a transverse wave in some other media) due to the vibration of molecules. In air, sound is transmitted as a pressure variation, between high and low pressure, with the rate of pressure variation from low, to high, to low again, determining the frequency. The degree of pressure variation (namely the difference between the high and the low) determines the amplitude.

A microphone captures sound waves, often by sensing the deflection caused by the wave on a thin membrane, transforming it proportionally to either voltage or current. The resulting electrical signal is normally then converted to a sequence of coded digital data using an analogue-to-digital converter (ADC). The most common format, pulse coded modulation, will be described in Section 5.1.1.

If this same sequence of coded data is fed through a compatible digital-to-analogue converter (DAC), through an amplifier to a loudspeaker, then a sound may be produced. In this case the voltage applied to the loudspeaker at every instant of time is proportional to the sample value from the computer being fed through the DAC. The voltage on the loudspeaker causes a cone to deflect in or out, and it is this cone which compresses (or rarifies) the air from instant to instant thus initiating a sound wave.

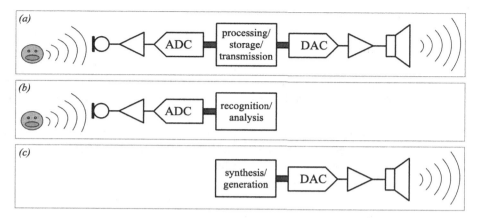

**Figure 1.1** Block diagram of three classes of digital audio system showing (a) a complete digital audio processing system comprising (from left to right) an input microphone, amplifier, ADC, digital system, DAC, amplifier and loudspeaker. Variations also exist for systems recognising audio or speech (b), and systems synthesising audio (c).

In fact the process, shown diagrammatically in Figure 1.1(a), identifies the major steps in any digital audio processing system. Audio, in this case speech in free air, is converted to an electrical signal by a microphone, amplified and probably filtered, before being converted into the digital domain by an ADC. Once in the digital domain, these signals can be processed, transmitted or stored in many ways, and indeed may be experimented upon using MATLAB. A reverse process will then convert the signals back into sound.

Connections to and from the processing/storage/transmission system of Figure 1.1 (which could be almost any digital system) may be either serial or parallel, with several standard options being available in either case. Optical and wireless variants are also increasingly popular.

Variations on this basic system, such as shown in Figure 1.1(b) and (c), use a subset of the components for analysis or synthesis of audio. Stereo systems would have two microphones and loudspeakers, and some systems may have many more of either. The very simple amplifier, ADC and DAC blocks in the diagram also hide some of the complexities that would be present in many systems – such as analogue filtering, automatic gain control, and so on, in addition to the type (class) of amplification provided.

Both ADC and DAC are also characterised in different ways: by their sampling rates, technology, signal-to-noise ratio, and dynamic range, usually determined by the number of bits that they output.

## 1.3    Sampling

Considering a sequence of audio samples, first of all we note that the time spacing between successive samples is almost always designed to be uniform. The frequency of this timing is referred to as the sampling rate, and in Figure 1.1 would be set through a periodic clock signal fed to the ADC and DAC, although there is no reason why both need the same sample rate – digital processing can be used to change sample rate. Using the well-known Nyquist criterion, the highest frequency that can be unambiguously represented by such a stream of samples is half of the sampling rate.

Samples themselves as delivered by ADC are generally fixed point with a resolution of 16 bits, although 20 bits and even up to 24 bits are found in high-end audio systems. Handling these on computer could utilise either fixed or floating point representation (fixed point meaning each sample is a scaled integer, while floating point allows fractional representation), with a general rule of thumb for reasonable quality being that 20 bits fixed point resolution is desirable for performing processing operations in a system with 16-bit input and output.

In the absence of other factors, the general rule is that an $n$ bit uniformly sampled digital audio signal will have a dynamic range (the ratio of the biggest amplitude that can be represented in the system to the smallest one) of, at best:

$$DR(dB) = 6.02 \times n. \tag{1.1}$$

For telephone-quality speech, resolutions as low as 8–12 bits are possible depending on the application. For GSM-type mobile phones, 14 bits is common. Telephone-quality, often referred to as toll-quality, is perfectly reasonable for vocal communications, but is not perceived as being of particularly high quality. For this reason, more modern vocal communication systems have tended to move beyond 8 bits sample resolution in practice.

Sample rates vary widely from 7.2 kHz or 8 kHz for telephone-quality audio to 44.1 kHz for CD-quality audio. Long-play style digital audio systems occasionally opt for 32 kHz, and high-quality systems use 48 kHz. A recent trend is to double this to 96 kHz. It is debatable whether a sampling rate of 96 kHz is at all useful to the human ear which can typically not resolve signals beyond about 18 kHz, apart from the rare listeners having *golden ears*.[1] However such systems may be more pet-friendly: dogs are reportedly able to hear up to 44 kHz and cats up to almost 80 kHz.

---

[1] The die-hard audio enthusiasts who prefer valve amplifiers, pay several years' salary for a pair of loudspeakers, and often claim they can hear above 20 kHz, are usually known as having *golden ears*.

---

**Infobox 1.1** Audio fidelity

Something to note is the inexactness of the entire conversion process: what you hear is a wave impinging on the eardrum, but what you obtain on the computer has travelled some way through air, possibly bounced past several obstructions, hit a microphone, vibrated a membrane, been converted to an electrical signal, amplified, and then sampled. Amplifiers add noise, distortion, and are not entirely linear. Microphones are usually far worse on all counts. Analogue-to-digital converters also suffer linearity errors, add noise, distortion, and introduce quantisation error due to the precision of their voltage sampling process. The result of all this is a computerised sequence of samples that may not be as closely related to the real-world sound as you might expect. Do not be surprised when high-precision analysis or measurements are unrepeatable due to noise, or if delicate changes made to a sampled audio signal are undetectable to the naked ear upon replay.

**Table 1.1.** Sampling characteristics of common applications.

| Application | Sample rate, resolution | Used how |
|---|---|---|
| telephony | 8 kHz,  8–12 bits | 64 kbps A-law or $\mu$-law |
| voice conferencing | 16 kHz, 14–16 bits | 64 kbps SB-ADPCB |
| mobile phone | 8 kHz, 14–16 bits | 13 kbps GSM |
| private mobile radio | 8 kHz, 12–16 bits | <5 kbps, e.g. TETRA |
| long-play audio | 32 kHz, 14–16 bits | minidisc, DAT, MP3 |
| CD audio | 44.1 kHz, 16–24 bits | stored on CDs |
| studio audio | 48 kHz, 16–24 bits | CD mastering |
| very high end | 96 kHz, 20–24 bits | for *golden ears* listening |

Sample rates and sampling precisions for several common applications, for humans at least, are summarised in Table 1.1.

## 1.4     Summary

Most of the technological detail related to the conversion and transmission process is outside the scope of this book, although some excellent resources covering this can be found in the bibliography. Generally, the audio processing specialist is fortunate enough to be able to work with digital audio without being too concerned with how it was captured, or how it will be replayed. Thus, we will confine our discussions throughout the remainder of this text primarily to the processing/storage/transmission, recognition/analysis and synthesis/generation blocks in Figure 1.1, ignoring the messy analogue detail.

Sound, as known to humans, has several attributes. These include time-domain attributes of duration, rhythm, attack and decay, but also frequency domain attributes of tone and pitch. Other, less well-defined attributes, include quality, timbre and tonality. Often, a sound wave conveys meaning: for example a fire alarm, the roar of a lion, the cry of a baby, a peal of thunder or a national anthem.

However, as we have seen, sound sampled by an ADC (at least the more common pulse coded modulation-based ADCs) is simply represented as a vector of samples, with each element in the vector representing the amplitude at that particular instant of time. The remainder of this book attempts to bridge the gap between such a vector of numbers representing audio, and an understanding or interpretation of the meaning of that audio.

## Bibliography

- *Principles of Computer Speech*
  I. H. Witten (Academic Press, 1982)
  This book provides a gentle and readable introduction to speech on computer, written in an accessible and engaging style. It is a little dated in the choice of technology presented, but the underlying principles discussed remain unchanged.

- *The Art of Electronics*
  P. Horowitz and W. Hill (Cambridge University Press, 2nd edition 1989)
  For those interested in the electronics of audio processing, whether digital or analogue, this book is a wonderful introduction. It is clearly written, absolutely packed full of excellent information (on almost any aspect of electronics), and a hugely informative text. Be aware though that its scope is large: with over 1000 pages, only a fraction is devoted to audio electronics issues.

- *Digital Signal Processing: A Practical Guide for Engineers and Scientists*
  S. W. Smith (Newnes, 2002)
  Also freely available from **www.dspguide.com**
  This excellent reference work is available in book form, or directly from the website above. The author has done a good job of covering most of the required elements of signal processing in a relatively easy-to-read way. In general the work lives up to the advertised role of being practically oriented. Overall, a huge amount of information is presented to the reader; however it may not be covered gradually enough for those without a signal processing background.

# 2 Basic audio processing

Audio is normal and best handled by MATLAB, when stored as a vector of samples, with each individual value being a double-precision floating point number. A sampled sound can be completely specified by the sequence of these numbers plus one other item of information: the sample rate. In general, the majority of digital audio systems differ from this in only one major respect, and that is they tend to store the sequence of samples as fixed-point numbers instead. This can be a complicating factor for those other systems, but an advantage to MATLAB users who have two less considerations to be concerned with when processing audio: namely overflow and underflow.

Any operation that MATLAB can perform on a vector can, in theory, be performed on stored audio. The audio vector can be loaded and saved in the same way as any other MATLAB variable, processed, added, plotted, and so on. However there are of course some special considerations when dealing with audio that need to be discussed within this chapter, as a foundation for the processing and analysis discussed in the later chapters.

This chapter begins with an overview of audio input and output in MATLAB, including recording and playback, before considering scaling issues, basic processing methods, then aspects of continuous analysis and processing. A section on visualisation covers the main time- and frequency-domain plotting techniques. Finally, methods of generating sounds and noise are given.

## 2.1 Handling audio in MATLAB

Given a high enough sample rate, the double precision vector has sufficient resolution for almost any type of processing that may need to be performed – meaning that one can usually safely ignore quantisation issues when in the MATLAB environment. However there are potential resolution and quantisation concerns when dealing with input to and output from MATLAB, since these will normally be in a fixed-point format. We shall thus discuss input and output: first, audio recording and playback, and then audio file handling in MATLAB.

## 2.1.1    Recording sound

Recording sound directly in MATLAB requires the user to specify the number of samples to record, the sample rate, number of channels and sample format. For example, to record a vector of double precision floating point samples on a computer with attached or integrated microphone, the following MATLAB command may be issued:

```
speech=wavrecord(16000,8000,1,'double');
```

This records 16 000 samples with a sample rate of 8 kHz, and places them into a 16 000 element vector named speech. The '1' argument specifies that the recording is mono rather than stereo. This command only works under Windows, so under Linux or MacOS it is best to use either the MATLAB audiorecorder() function, or use a separate audio application to record audio (such as the excellent open source audacity tool), saving the recorded sound as an audio file, to be loaded into MATLAB as we shall see shortly.

---

**Infobox 2.1** Audio file formats

**Wave**: The wave file format is usually identified by the file extension .wav, and actually can hold many different types of audio data identified by a header field at the beginning of the file. Most importantly, the sampling rate, number of channels and number of bits in each sample are also specified. This makes the format very easy to use compared to other formats that do not specify such information, and thankfully this format is recognised by MATLAB. Normally for audio work, the wave file would contain PCM data, with a single channel (mono), and 16 bits per sample. Sample rate could vary from 8000 Hz up to 48 000 Hz. Some older PC sound cards are limited in the sample rates they support, but 8000 Hz and 44 100 Hz are always supported. 16 000 Hz, 24 000 Hz, 32 000 Hz and 48 000 Hz are also reasonably common.

**PCM** and **RAW** hold streams of pulse coded modulation data with no headers or gaps. They are assumed to be single channel (mono) but the sample rate and number of bits per sample are not specified in the file – the audio researcher must remember what these are for each .pcm or .raw file that he or she keeps. These can be read from and written to by MATLAB, but are not supported as a distinctive audio file. However these have historically been the formats of choice for audio researchers, probably because research software written in C, C++ and other languages can most easily handle this format.

**A-law** and $\mu$-**law** are logarithmically compressed audio samples in byte format. Each byte represents something like 12 bits in equivalent linear PCM format. This is commonly used in telecommunications where the sample rate is 8 kHz. Again, however, the .au file extension (which is common on UNIX machines, and supported under Linux) does not contain any information on sample rate, so the audio researcher must remember this. MATLAB does support this format natively.

**Other** formats include those for compressed music such as MP3 (see Infobox: Music file formats on page 11), MP4, specialised musical instrument formats such as MIDI (musical instrument digital interface) and several hundred different proprietary audio formats.

---

If using the audiorecorder() function, the procedure is first to create an audio recorder object, specifying sample rate, sample precision in bits, and number of channels, then to begin recording:

```
aro=audiorecorder(16000,16,1);
record(aro);
```

At this point, the microphone is actively recording. When finished, stop the recording and try to play back the audio:

```
stop(aro);
play(aro);
```

To convert the stored recording into the more usual vector of audio, it is necessary to use the getaudiodata() command:

```
speech=getaudiodata(aro, 'double');
```

Other commands, including pause() and resume(), may be issued during recording to control the process, with the entire recording and playback operating as background commands, making these a good choice when building interactive speech experiments.

## 2.1.2    Storing and replaying sound

In the example given above, the 'speech' vector consists of double precision samples, but was recorded with 16-bit precision. The maximum representable range of values in 16-bit format is between $-32\,768$ and $+32\,767$, but when converted to double precision is scaled to lie with a range of $+/-1.0$, and in fact this would be the most universal scaling within MATLAB so we will use this wherever possible. In this format, a recorded sample with integer value $32\,767$ would be stored with a floating point value of $+1.0$, and a recorded sample with integer value $-32\,768$ would be stored with a floating point value of $-1.0$.

Replaying a vector of sound stored in floating point format is also easy:

```
sound(speech, 8000);
```

It is necessary to specify only the sound vector by name and the sample rate (8 kHz in this case, or whatever was used during recording). If you have a microphone and speakers connected to your PC, you can play with these commands a little. Try recording a simple sentence and then increasing or reducing the sample rate by 50% to hear the changes that result on playback.

Sometimes processing or other operations carried out on an audio vector will result in samples having a value greater than $+/-1.0$, or in very small values. When replayed using sound(), this would result in clipping, or inaudible playback respectively. In such cases, an alternative command will automatically scale the audio vector prior to playback based upon the maximum amplitude element in the audio vector:

```
soundsc(speech, 8000);
```

This command scales in both directions so that a vector that is too quiet will be amplified, and one that is too large will be attenuated. Of course we could accomplish something similar by scaling the audio vector ourselves:

```
sound(speech/max(abs(speech)), 8000);
```

It should also be noted that MATLAB is often used to develop audio algorithms that will be later ported to a fixed-point computational architecture, such as an integer DSP (digital signal processor), or a microcontroller. In these cases it can be important to ensure that the techniques developed are compatible with integer arithmetic instead of floating point arithmetic. It is therefore useful to know that changing the 'double' specified in the use of the `wavrecord()` and `getaudio()` functions above to an 'int16' will produce an audio recording vector of integer values scaled between $-32\,768$ and $+32\,767$.

The audio input and output commands we have looked at here will form the bedrock of much of the process of audio experimentation with MATLAB: graphs and spectrograms (a plot of frequency against time) can show only so much, but even many experienced audio researchers cannot repeatedly recognise words by looking at plots! Perfectly audible sound, processed in some small way, might result in highly corrupt audio that plots alone will not reveal. The human ear is a marvel of engineering that has been designed for exactly the task of listening, so there is no reason to assume that the eye can perform equally as well at judging visualised sounds. Plots can occasionally be an excellent method of visualising or interpreting sound, but often listening is better.

A time-domain plot of a sound sample is easy in MATLAB:

```
plot(speech);
```

although sometimes it is preferred for the $x$-axis to display time in seconds:

```
plot( [ 1: size(speech) ]  / 8000, speech);
```

where again the sample rate (in this case 8 kHz) needs to be specified.

## 2.1.3     Audio file handling

In the audio research field, sound files are often stored in a raw PCM (pulse coded modulation) format. That means the file consists of sample values only – with no reference to sample rate, precision, number of channels, and so on. Also, there is a potential endian problem for samples greater than 8 bits in size if they have been handled or recorded by a different computer type.

To read raw PCM sound into MATLAB, we can use the general purpose `fread()` function, which has arguments that specify the data precision of the values to read in from a binary file, and also the endianess (see Infobox: The endian problem on page 14).

First open the file to be read by name:

```
fid=fopen('recording.pcm', 'r');
```

Next, read in the entire file, in one go, into a vector:

```
speech=fread(fid , inf , 'int16' , 0, 'ieee-le');
```

This would now have read in an entire file ('inf' or infinite values) of 16-bit integers. The format is IEEE little endian, which is what a PC tends to use. Alternatively (but rarely these days) we could have done:

```
speech=fread(fid , inf , 'uint16' , 0, 'ieee-be');
```

which would read in an entire file of unsigned 16-bit integers, in big endian format (such as a large UNIX mainframe might use).

Finally it is good practice to close the file we had opened once we have finished reading from it:

```
fclose(fid);
```

It is also useful to know how to save and load arrays within MATLAB. Using a built-in MATLAB binary format, an array of speech, can be saved to disc using the `save` command, and loaded using the `load` command. The normal filename extension for the stored file is '.mat'.

---

**Infobox 2.2** Music file formats

**MP3**, represented by the file extension .mp3, is a standard compressed file format invented by the Fraunhofer Institute in Germany. It has taken the world by storm: there is probably more audio in this format than in any other. The success of MP3, actually MPEG (Motion Pictures Expert Group) version 1 layer 3, has spawned numerous look-alikes and copies. These range from traditional technology leaders such as Apple, to traditional technology followers such as Microsoft.

**Ogg Vorbis**, one notable effort is this strangely named format which is comparable in functionality to MP3, but not compatible with it: it is solely designed to be an open replacement for MP3, presumably for anyone who does not wish to pay licence fees or royalties to the Fraunhofer Institute. As such it has seen widespread adoption worldwide. However one thing is certain, and that is the world is currently locked into a battle between these formats, with a large and growing economic reason for ensuring format dominance.

Luckily for the audio researcher, compressed file formats tend to destroy audio features, and thus are not really suitable for storage of speech and audio for many research purposes, thus we can stay out of the controversy and confine ourselves to PCM, RAW and Wave file formats.

---

For example, two vectors in the MATLAB workspace called `speech` and `speech2` could be saved to file 'myspeech.mat' in the current directory like this:

```
save   myspeech.mat   speech speech2
```

Later, the saved arrays can be reloaded into another session of MATLAB by issuing the command:

```
load   myspeech.mat
```

There will then be two new arrays imported to the MATLAB workspace called `speech` and `speech2`. Unlike with the `fread()` command used previously, in this case the name of the stored arrays is specified in the stored file.

## 2.1.4    Audio conversion problems

Given the issue of unknown resolution, number of channels, sample rate and endianess, it is probably useful to listen to any sound after it is imported to check it was converted correctly (but please learn from an experienced audio researcher – always turn the volume control right down the first time that you replay any sound: pops, squeaks and whistles, at painfully high volume levels, are a constant threat when processing audio, and have surprised many of us). You could also plot the waveform, and may sometimes spot common problems from a visual examination. Figure 2.1 shows an audio recording plotted directly, and quantised to an unsigned 8-bit range on the top of the figure. On the bottom, the same sound is plotted with incorrect byte ordering (in this case where each 16-bit sample has been treated as a big-endian number rather than a little-endian number), and as an absolute unsigned number. Note that all of these examples, when heard by ear, result in understandable speech – even the incorrectly byte ordered replay (it is easy to verify this, try the MATLAB `swapbytes()` function in conjunction with `soundsc()`).

Other problem areas to look for are recordings that are either twice as long, or half as long as they should be. This may indicate an 8-bit array being treated as 16-bit numbers, or a 16-bit array being treated as doubles.

As mentioned previously, the ear is often the best discriminator of sound problems. If you specify too high a sample rate when replaying sound, the audio will sound squeaky, and will sound s-l-o-w if the sample rate is too low. Incorrect endianess will probably cause significant amounts of noise, and getting unsigned/signed mixed up will result in noise-corrupted speech (especially with loud sounds). Having specified an incorrect precision when loading a file (such as reading a logarithmic 8-bit file as a 16-bit linear) will often result in a sound playback that is noisy but recognisable.

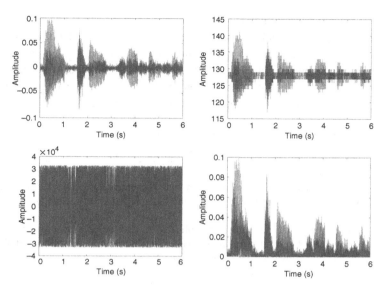

**Figure 2.1** Four plots of an audio waveform shown unaltered on top left, correctly quantised to unsigned 8-bit number format on top right, with incorrect byte ordering on bottom left and converted as an absolute unsigned number on the bottom right.

## 2.2 Normalisation

There is one final step to basic audio handling, and that is normalising the sample vector. If you notice when we discussed replaying sounds earlier, we sometimes had to normalise first to prevent clipping:

```
sound(speech/max(abs(speech)), 8000);
```

Just to recap, the reason for this is that MATLAB expects each element in the sound vector to be scaled into a range of between $-1.0$ and $+1.0$. However the audio that we imported was probably 16-bit signed linear fixed point format, having a scale several thousand times larger than this. Even if the audio was presented in the correct range, processing it often causes samples to extend beyond their original range.

It is therefore good practice to comply with MATLAB's expectations and scale audio being processed to the expected range after processing, unless it is important to maintain bit exactness with some digital system, or to investigate quantisation issues themselves. In general we can scale in two ways:

**Absolute scaling** considers the format that the audio was captured in, and scales relative to that (so we would divide each element in the input vector by the biggest value in that representation: 32768 for 16-bit signed linear).

**Relative scaling** scales relative to the largest value in the sample vector. This is the method we used when playing back audio earlier.

In general the choice of scaling method depends on whether the absolute amplitude of the original sound is important; for example if you are handling many music recordings for a performance, then you would want to preserve some pieces as being quieter than others, so you would use absolute scaling. On the other hand, if you wanted to detect the pitch in recorded speech, you might use relative scaling since it probably doesn't matter how loud the speech was originally as long as the pitch is evident.

---

**Infobox 2.3** The endian problem

The two competing formats are big and little endian. Big endian means the most significant byte is presented/stored first, and is used by computers such as Sun and HP workstations. Little endian means that the least significant byte is presented/stored first, as used by the Intel and AMD processors inside most desktop PCs and laptops. Some processors (such as the ARM7) allow for switchable endianess.

Unfortunately, endianess is complicated by the variable access-width of modern computers. When everything was byte-wide it was easier, but now there is an added dimension of difficulty. Given an unknown system, it is probably easier to check if it is little endian, and if not classify it as big endian rather than working the other way around.

**Example**

Given a 32-bit audio sample stored in a byte-wide file system (a very common scenario), with the stored word being made up of least significant byte (LSB), second most significant byte (B1), third most significant byte (B2) and most significant byte (MSB). Does the following diagram show a little or big-endian representation?

```
4
3    MSB
2    B2
1    B1
0    LSB
     7   0
```

In the diagram, the storage address (in bytes) is given on the left, and the bit position shown below. In this case the bit positions are not really important.

Checking for little endian first, we identify the lowest byte-wide address, and count upwards, looking at the order in which the stored bytes are arranged. In this case the lowest address is 0 and the lowest byte starts at bit 0. The next byte up holds B1, and so on. So counting the contents from lowest byte address upwards, we get {LSB, B1, B2, MSB}. Since this DOES follow least-to-most it must be little endian. By contrast, the following diagram shows a big-endian representation of a 16-bit sample:

```
2
1    LSB
0    MSB
     7   0
```

These days, by using the wave file format, the endianess is taken care of. It is also irrelevant with byte-wide formats such as 8-bit samples or A-law samples; however problems do arise when handling raw PCM audio files that are sampled at 16-bit.

## 2.3 Audio processing

Probably the most important consideration with handling audio is the question of whether the sound vector can be processed in one go, or whether it needs to be split up and processed in pieces. In a real sound processing system, these questions would usually revolve around when an output was required – a system able to split incoming audio and process piece-by-piece would start to produce an output earlier than if output was generated only at the end of an input recording. Other issues would include the degree of processing power required (such as determining whether all processing need to be performed in one 'lump' at the end of an input, or could it be spread over the input time?).

Within MATLAB, short recordings may be processed in their entirety, but longer recordings may well consume all of the memory within the computer. More importantly where audio characteristics evolve over time, such as in speech, it is often useful to split speech into either equal-sized regions or regions of similar characteristic. This splitting is called segmentation and will be discussed further in Section 2.4.

Many time domain operations, such as a simple digital filter, can be applied to an audio vector in a single pass irrespective of its size (and whether or not it is has been segmented). For example an FIR (finite impulse response) filter is achieved with:

```
y=filter(b, 1, x);
```

where $b$ is a vector holding the filter coefficients. The $x$ and $y$ vectors are equal length vectors of input and output samples respectively, with each sample of $y$ being calculated from the difference equation:

$$y(n) = b(1) \times x(n) + b(2) \times x(n-1) + b(3) \times x(n-2) + \cdots$$
$$+ b(m+1) \times x(n-m). \tag{2.1}$$

An IIR or pole-zero filter is achieved with:

```
y=filter(b, a, x);
```

both a and b are vectors holding the filter coefficients. $x$ and $y$ are again equal length vectors of input and output samples respectively, with each sample of $y$ being calculated from the difference equation:

$$y(n) = b(1) \times x(n) + b(2) \times x(n-1) + b(3) \times x(n-2) + \cdots$$
$$+ b(m+1) \times x(n-m)$$
$$- a(2) \times y(n-1) - a(3) \times y(n-2) \cdots \tag{2.2}$$
$$- a(m+1) \times y(n-m).$$

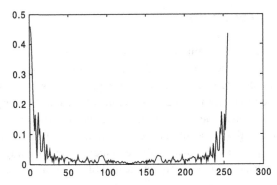

**Figure 2.2** Absolute FFT plot for audio spectrum, with frequency index along the x-axis and amplitude along the y-axis.

Frequency domain operations, by contrast, usually require the audio to be first converted to the frequency domain by use of a Fourier transform or similar, such as the Fast Fourier Transform (FFT):

```
a_spec=fft(a_vector);
```

or more generally when the audio vector length is not a power of two, it is possible to zero-pad (or truncate) the audio vector to fill the size of FFT specified, as the following illustrates for a 256-element FFT:

```
a_spec=fft(a_vector, 256);
```

The pertinent question which arises is 'how big is that transform?' Which is also asking 'what frequency resolution is required?' The reader should already be aware that the number of frequency bins in the FFT output is based on the number of samples given as input. This will be explored further in Section 2.5.2, but for the present, suffice it to say that a convenient power-of-two size is normally chosen for frequency vector length. Another way of achieving this is:

```
a_spec=fft(a_vector(1:256));
```

In MATLAB, the resultant frequency-domain vector will be complex. Plotting the absolute value of the vector provides a double-sided frequency representation shown in Figure 2.2 and plotted using:

```
plot(abs(a_spec));
```

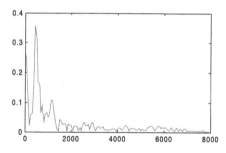

**Figure 2.3** Single-sided absolute FFT plot for the same audio spectrum as shown in Figure 2.2.

In this unusual plot, the frequency axis (if scaled correctly), would start at 0, progress to the Nyquist frequency at the centre point, and then decrease to 0 at the far right. Both positive and negative frequencies are shown – something which is not particularly useful. In fact MATLAB differs in this way (for historical reasons) from many other FFT libraries in use for C and FORTRAN programmers. We can produce a more standard plot with the low frequencies in the centre of the plot using:

```
plot(abs(fftshift(a_spec)));
```

However in audio processing we tend to plot the single-sided spectrum – and give it more useful axes. Plotting the same spectrum with variables `Fs=8000` and `Ns=256` describing the original sample rate and size of the FFT respectively, then a better plot would be achieved with:

```
plot( [1 : 2*Fs/Ns : Fs],  abs(a_spec(1:Ns/2)), 'r');
```

This plots the spectrum as shown in Figure 2.3, which is clearly a more useful, and physically relevant representation, with the 'r' argument to `plot()` meaning the plotted line is coloured red on a colour display.

Of course when performing audio processing, some form of analysis would typically be performed on the frequency vector that results from an FFT. This is all well and good, but what if the audio vector contains many more than 256 samples? The answer is that the longer vector will be split (or segmented) into several 256-sample frames, and each frame handled separately.

Segmentation is needed not only because 256 (say) is a convenient size, but when any of the following are true:

**1.** The audio is continuous (i.e. you can't wait for a final sample to arrive before beginning processing).

2. The nature of the audio signal is continually changing, or short-term features are important (i.e. a tone of steadily increasing frequency may be observed by a smaller Fourier transform snapshot but would average out to white noise if the entire sweep is analysed at once).
3. The processing applied to each block scales nonlinearly in complexity (i.e. a block twice as big would be four or even eight times more difficult to process).
4. In an implementation, memory space is limited (very common).
5. It is desirable to spread processing over a longer time period, rather than performing it all at the end of a recording.
6. Latency (the delay between the first sample in the block and the analysis output) is to be minimised – a common requirement for voice communication systems.

## 2.4        Segmentation

Segmentation into frames is a basic necessity for much audio processing as mentioned above in Section 2.3, but the process of segmentation does have its own fair share of problems.

Consider an audio 'feature'. By that I mean some type of sound that is contained within a vector of samples. Now when that vector is analysed it might happen that the feature is split into two: half appears in one audio frame, and the other half in another frame. The complete feature does not appear in any analysis window, and may have effectively been hidden. In this way, features that are lucky enough to fall in the centre of a frame are emphasised at the expense of features which are chopped in half. When windowing is considered (see Section 2.4.2), this problem is exacerbated further since audio at the extreme ends of an analysis frame will be de-emphasised further. The solution to the lost-feature problem is to overlap frames.

### 2.4.1        Overlap

Overlap means that instead of straightforward segmentation of the audio vector into sequential frames, each new frame is made to contain a part of the previous frame and part of the next frame. Overlapping ensures that audio features occurring at a discontinuity are at least considered whole in the subsequent, overlapped, frame.

The degree of overlap (usually expressed as a percentage) describes the amount of the previous frame that is repeated in the following frame. Overlaps of 25% and 50% are common.

Figure 2.4 illustrates the process for 50% overlap. The top graph shows a plot of a waveform over time. This is then split into a sequence of overlapping frames, numbered 0 to 6. Every feature in the original signal can now be found repeated in two frames.

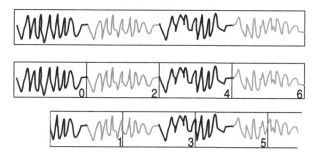

**Figure 2.4** Illustration of an original audio recording (the upper waveform) divided into two offset sequences of analysis windows (two lower waveforms) with 50% overlapping frames so that no short-term auditory feature is obscured by straddling the boundary between analysis windows: it will appear unbroken in at least one of the two analysis streams.

Readers with experience of implementation might immediately notice one disadvantage with overlap: namely that we end up with more audio: a 50% overlap doubles the number of frames that need to be processed.

Both segmentation and overlap also cause problems with processing, as opposed to analysis. Note that analysis is simply a one-way process where audio goes into an analyser, and some measure comes out. Processing differs in that audio enters the processing system, and then (presumably different) audio comes out again.

So now imagine each frame of audio being processed in some way. For example, if each frame is scaled with respect to its average value, sometimes referred to as autozeroing, then the frames are stuck back together to produce averaged output audio.

First of all is the question of how to cope with the overlap. The frames cannot simply be concatenated because there would then be twice as many samples, and adding them together won't work either.

Even with no overlap Figure 2.5 illustrates another problem if frames have been split, processed and then rejoined: there is a discontinuity between neighbouring frames. Something like this is very easily audible to the ear which perceives it as a clicking sound. In fact it turns out that almost any type of non-trivial processing will result in such discontinuities.

## 2.4.2    Windowing

The work-around for many of the segmentation and overlap problems is windowing prior to reassembly. A judicious choice of window alleviates the joining together of overlapped, processed frames (see, for example, Chapter 18 of [1]). Windowing is also a prerequisite for frequency analysis to prevent edge effects – Gibbs phenomena – by the FFT [2].

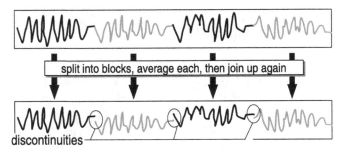

**Figure 2.5** Illustration of an original audio recording (upper waveform) being split into equal length analysis windows, each of which is normalised, and then rejoined (lower waveform). The rejoined audio exhibits discontinuities between frames which would result in significant audio distortion.

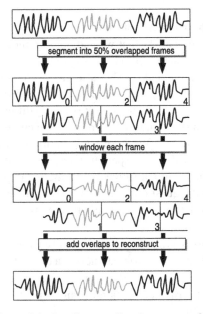

**Figure 2.6** Illustration of an original audio recording (upper waveform) being split into 50% overlapped analysis frames, each of which are windowed before being summed together to reconstruct output audio which does not exhibit discontinuities (lower waveform).

Figure 2.6 illustrates some audio being segmented into 50% overlapping frames, windowed, and then reconstructed.

There are many predefined window types in common use, and with different characteristics (that are adequately explained in most signal processing textbooks). If in doubt, it's probably best to use a Hamming window (although it does not tail off quite to zero

**Table 2.1.** Common window functions.

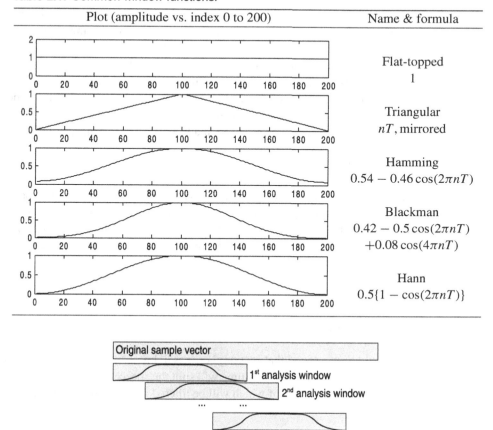

| Plot (amplitude vs. index 0 to 200) | Name & formula |
|---|---|
| | Flat-topped<br>1 |
| | Triangular<br>$nT$, mirrored |
| | Hamming<br>$0.54 - 0.46\cos(2\pi nT)$ |
| | Blackman<br>$0.42 - 0.5\cos(2\pi nT)$<br>$+0.08\cos(4\pi nT)$ |
| | Hann<br>$0.5\{1 - \cos(2\pi nT)\}$ |

**Figure 2.7** Illustration of overlap analysis employing a custom window shape.

at either end). Several examples of the more common window functions are given in Table 2.1.

Many other windows are available, and any reader that gets bored is encouraged to invent his own. In addition, the existing windows can be skewed one way or another within an analysis frame, or split into two and a flat-topped section inserted, which is especially useful when using 25% overlap. See Figure 2.7 for an example.

## 2.4.3 Continuous filtering

In Sections 2.4.1 and 2.4.2 we discussed how to segment-window-overlap speech in order to 'smooth out' transitions between different processing domains.

Now, in general, this may only be required when the processing performed on neighbouring analysis frames is different. If, for example, the *same* digital filter acts on two neighbouring analysis frames, then is there any reason why there should be any discontinuity between them? Well it turns out that yes, there may well be a discontinuity unless we carefully consider the way we perform the filtering.

Let us illustrate this effect in MATLAB with some speech which we can filter and replay. First, record a few seconds of speech at an 8 kHz sample rate as described in Section 2.1. We shall assume this is in an array named s.

As we know this could be replayed and listened to very simply:

```
soundsc(s);
```

Next we will define a small digital filter to act upon the speech. For the purpose of testing, we can define a simple transfer function $h = (1 - 0.9375z^{-1})$, which is actually one we will encounter later in Chapter 5, used for the pre-emphasis of speech. In the following MATLAB code, we will create the filter, apply to the entire array of speech, then listen to the result, in vector y:

```
h=[1, -0.9375];
y=filter(h, 1, s);
soundsc(y);
```

We should hear a sound that is 'tinny' or 'nasal' but is otherwise clear speech. Now we will repeat the filtering, but this time will split the speech recording into a succession of 240 sample frames, each of which will be filtered individually, and then reassembled:

```
w=240;
n=floor(length(s)/w);
for k=1:n
   seg=s(1+(k-1)*w:k*w);
   segf=filter(h, 1, seg);
   outsp(1+(k-1)*w:k*w)=segf;
end
soundsc(outsp);
```

The output should still sound similar, but will most likely be marred by a succession of audible clicks. Interestingly, if we were to plot the filtered output signals y and outsp, they would probably appear indistinguishable. It is only using methods like the spectrogram (see Section 2.6) that these clicks or discontinuities would become visible.

So now two questions remain: firstly, what causes the discontinuities, and secondly how can we prevent them from happening? The cause may be immediately obvious if one considers exactly what the filtering process is doing. In effect, an FIR filter is describing a process where 'the current output sample is a sum of the current input sample plus the previous $P$ input samples, each scaled by some coefficients', where $P$ is the filter order. Thus in the first-order filter shown, the current output sample will equal the current input sample minus 0.9375 times the previous input sample:

$$y[n] = s[n] - 0.9375 \times s[n-1] \tag{2.3}$$

where the index $n$ ranges from 0 to the length of the speech vector.[1] The problem comes at the start of each frame, with index $n = 0$. Consider the case of the second frame to be filtered. When calculating $y[n]$ at the start of that frame, index $n = 0$ so $y[0]$ depends upon $s[-1]$, which will itself be zero. However, when calculating the same output value without having segmented the speech, the index would not be zero but 240 in this case. Given that $y[240] = s[240] - 0.9375 \times s[239]$, and since $s[239]$ is known, the output would differ to the calculation in the segmented frame.

Now this is only a single tap filter. Imagine a filter having more taps: the number of samples affected at the start of each new frame would be more, and thus the size of discontinuities is likely to be greater. In a tenth-order speech coder, this effect, if present, would very significantly degrade the quality of output speech.

These past coefficients within a digital filter are termed the filter *history*, or its internal state. When filtering an array, the internal state is automatically updated sample-by-sample. However when filtering the beginning of an array, the internal state needs to be specified, otherwise it will default to zero. In the segmented speech case, the filter history is actually being reset to zero at the boundary of each frame – no wonder the output is distorted.

MATLAB actually provides a convenient way to set and store internal history using the filter() command. We will illustrate this by editing and repeating the segmented example above:

```
w=240;
hst=[];
n=floor(length(s)/w);
for k=1:n
   seg=s(1+(k-1)*w:k*w);
   [segf, hst]=filter(h, 1, seg, hst);
   outsp2(1+(k-1)*w:k*w)=segf;
end
soundsc(outsp2);
```

In the updated code, we are now passing in a history array to the filter() command, and each time we perform the filter we are storing the resulting internal history. By specifying the filter history at the input to each subsequent frame, we are setting those internal coefficients to the state that they were in when the filter completed its operation

---

[1] Remember in MATLAB, the first index to an array is 1 rather than 0. Although we will use MATLAB to illustrate the effect, we will use the common engineering approach of indexing from zero in our discussion and equations.

on the previous frame. We thus ensure a smooth filtering operation across frames. The resulting output speech should be free of the clicks and discontinuities evident with the previous code.

## 2.5    Analysis window sizing

We have discussed at the end of Section 2.3 several motivations for splitting audio into segments for processing, but we did not consider how big those segments, frames or analysis windows, should be. Generally, most audio algorithms (and definitely MATLAB-based processing) will operate more efficiently on larger blocks of data. There would therefore be a natural tendency toward using larger analysis frames, tempered by issues such as latency which is a critical consideration in telephony processing and similar applications.

Another major reason for limiting analysis window size is where the characteristics of a signal change during that analysis window. This is perhaps best illustrated in the Infobox Visualisation of signals on page 32 where a complex frequency-time pattern is present, but an analysis window which is large enough to span across that pattern will hide the detail when an FFT is performed.

There are two important points to be explained here. The first is that of *signal stationarity* and the second is *time-frequency resolution*. We will consider each in turn.

### 2.5.1    Signal stationarity

Most signals requiring analysis are continually changing. A single sustained note played on a musical instrument is stationary, but quite clearly when one note is replaced by the next one, the signal characteristics have changed in some way (at least in frequency, but possibly also in amplitude, tone, timbre, and so on).

For an application analysing recorded music to determine which note is currently being played, it would make sense to segment the recording roughly into analysis windows of length equal to the duration of a single note, or less. For each analysis window we could perform an FFT, and look for peaks in the spectrum. However if we analysed longer duration windows, we may end up performing an FFT that spans across two notes, and be unable to determine which is either note. At very least we would have a confused 'picture' of the sound being analysed – just as the example FFT in the Infobox did not reveal the full detail of the sound being analysed.

More importantly, the theory that gives rise to the FFT assumes that the frequency components of the signal are unchanging across the analysis window of interest. Any deviation from this assumption would result in an inaccurate determination of the frequency components.

These points together reveal the importance of ensuring that an analysis window leading to FFT be sized so that the signal is stationary across the period of analysis. In practice many audio signals do not tend to remain stationary for long, and thus smaller analysis windows are necessary to capture the rapidly changing details.

In speech analysis, as will be described in Chapter 3, many of the muscle movements which cause speech sounds are relatively slow moving, resulting in speech which slowly changes its spectral characteristics. A useful rule of thumb is that the speech signal can be assumed to be stationary, in fact pseudo-stationary, over a period of about 20–30 ms. Thus speech analysis typically requires segmentation into 20 ms long frames [3].

The stationarity requirement also extends to linear prediction (Section 5.2.1) and many other forms of analysis. When used, each must be carefully matched against the known characteristics of the audio signals which are to be handled.

## 2.5.2   Time-frequency resolution

Moving back to the FFT, the output frequency vector, from an $N$-sample FFT of audio sampled at $Fs$ Hz, contains $N/2 + 1$ positive frequency bins. Each bin collects the energy from a small range of frequencies in the original signal. The bin width is related to both the sampling rate and to the number of samples being analysed, $Fs/N$. Put another way, this bin width is equal to the reciprocal of the time span encompassed by the analysis window.

It therefore makes sense that, in order to achieve a higher frequency resolution, we need to collect a longer duration of samples. However for rapidly changing signals, collecting more of them means we might end up missing some time-domain features as discussed in Section 2.5.1 and Infobox Visualisation of signals on page 32.

So there is a basic uncertainty principle operating here: a single FFT can trade off between higher frequency resolution (more samples) or higher time resolution (fewer samples) but cannot do both simultaneously. Solutions vary with the requirements of the problem, but there are several frequency estimation alternatives to the FFT, and it may often be possible to perform two FFTs, over long and short analysis windows, respectively. Later in Section 6.2.2 we will describe more computationally intensive methods of attempting to satisfy both the demand of high frequency resolution and of high time resolution.

## 2.6   Visualisation

**Plot of waveform:** this is the easiest and most basic method of visualisation, and can be very useful, especially for quickly scanning very long recordings. But beware that most of the information is hidden and what you expect to hear after viewing a waveform doesn't always tie up with what you hear.

**Frequency spectrum:** again this was mentioned previously, and is a basic and well-used tool. As long as it is applied correctly, and to the correct section of audio: it is very easy to obtain a spectrum from the wrong section of a recording, thus missing an important feature. If an entire long recording needs to be visualised in frequency terms, then use the following:

**Short-time Fourier transform (STFT):** is a sliding-window narrow Fourier transform that is repeated sequentially over a long vector of samples, performing

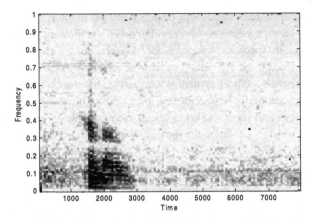

**Figure 2.8** A spectrogram of human speech, plotting normalised frequency against time (in milliseconds). Louder sound components are shown with a darker grey shade, and lower amplitudes with a lighter shade.

time-frequency signal decomposition or analysis. This results in a time sequence of individual spectra, which can be plotted against time, either in *x-y-z* graph, or as a spectrogram (which is actually the magnitude squared of the transformed data):

$$\text{spectrogram } x(x) = |X(\tau, \omega)|^2.$$

In MATLAB this is easy to perform with the specgram() function, which is however unfortunately destined to be removed in a future version of MATLAB. The replacement, spectrogram(), is available in the signal processing toolkit and does much the same thing – having many options regarding analysis window size, overlap and number of sample bins.

A spectrogram is essentially a set of STFT plotted as frequency against time with the intensity (*z*-axis) given as a greyscale, or colour pixel. For speech analysis, the spectrogram is an excellent method of visualising speech structure and how it changes over time.

An example of using MATLAB for plotting a spectrogram is shown in Figure 2.8, plotted in greyscale. Some audio researchers prefer to plot their spectrograms in colour. It is really just a matter of personal preference.

## 2.6.1    A brief note on axes

The horizontal axis of an FFT plot, or spectrum, is traditionally used to represent frequency, whilst the vertical axis would display amplitude. A spectrogram, by contrast, plots time along the horizontal axis, frequency on the vertical and amplitude on the *z*-axis (as colour or greyscale).

Due to the large dynamic range of audio signals, it is common to plot the logarithm of absolute amplitude, rather than the amplitude directly; thus the spectrum will often be plotted as a power spectrum, using $20 \times \log_{10}(\text{spectrum})$, or with the MATLAB:

```
semilogy(spectrum);
```

The $x$-axis, showing frequency, is generally plotted linearly. The labelling of this axis defaults to the index number of the bins in the FFT output spectrum – not particularly useful in most cases. A far better approach is to specify this yourself either directly in Hz, scaled between 0 and 1 (DC to Nyquist frequency), or in radians between 0 and $\pi$.

```
res=pi/size(spectrum);
semilogy(res:res:pi,  spectrum);
```

This radian measure, normally denoted by the independent frequency variable being written as $\omega$, represents $2\pi$ as the sampling frequency, 0 as DC and thus $\pi$ to be the Nyquist frequency. It is referred to as *angular frequency* or occasionally *natural frequency*, and considers frequencies to be arranged around a circle. The notation is useful when dealing with systems that are over- or undersampled, but apart from this, it is more consistent mathematically because it means equations can be derived to describe a sampled system that do not depend on the absolute value of sample rate.

## 2.6.2    Other visualisation methods

As you may expect, many other more involved visualisations exist, and which have evolved as being particularly suited for viewing certain features. One of the most useful for speech and audio work is the linear prediction coefficient spectral plot that will be described in Section 5.2.1. Here, on the other hand, two very general and useful methods are presented – namely the correlogram and the cepstrum.

### 2.6.2.1    Correlogram

A correlogram is a plot of the autocorrelation of a signal. Correlation is the process by which two signals are compared for similarities that may exist between them either at the present time or in the past (however much past data are available). Mathematically it is relatively simple. We will start with an equation, defining the cross-correlation between two vectors $x$ and $y$ performed at time $t$, and calculating for the past $k$ time instants, shown in Equation (2.4):

$$c_{x,y}[t] = \sum_k x[k]y[t-k]. \tag{2.4}$$

In MATLAB such an analysis is performed using the `xcorr` function over the length of the shortest of the two vectors being analysed:

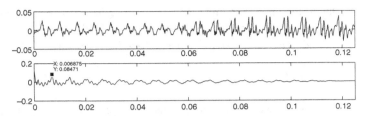

**Figure 2.9** Plot of a segment of relatively periodic speech (above) and its autocorrelation analysis (below), both plotted as amplitude against time, with the first correlation peak highlighted.

```
[c,lags]=xcorr(x,y);
```

If only a single argument is provided to xcorr then the analysis results in an auto-correlation, which compares the single vector against each possible time-shifted version of itself. The comparison at time shift $k$ is the sum of the product of each element of the input vector, with the element $k$ positions previous to it. Using this as a tool, we can look for periodicities in the vector being analysed.

As an example, if the input vector contains two peaks, then the autocorrelation output will be large at time shifts where one peak ends up being multiplied by the other peak. In that case, the time shift $k$ corresponding to the largest output value would be equal to the separation between the two peaks, in samples. Where the vectors being analysed are more complicated, autocorrelation analysis is a good way of looking for periods in the signal that might be hidden to the naked eye.

As an example, assume that we have a short recording of speech in MATLAB, imaginatively called speech. By plotting this, we notice that there is a section in there with a fairly regular structure, perhaps from a voiced segment of audio. For the illustrative purposes, assume this lies between vector elements 9000 and 10 000 in our recording. We are going to 'cut out' that section for autocorrelation analysis:

```
segment=speech(9000:10000);
```

Next we perform an autocorrelation analysis and plot both the speech segment and the resulting correlogram. In this example, the plotting commands are also reproduced for reference:

```
[c,lags]=xcorr(segment);
subplot(2,1,1);
plot([0:1:1000]/8000,segment);
axis([0, 0.125. -0.05, 0.05]);
subplot(2,1,2);
plot(lags(1001:2001)/8000,c(1001:2001));
axis([0, 0.125. -0.2, 0.2]);
```

The argument to the axis commands to define the $y$-dimension (from $-0.05$ to $0.05$ and from $-0.2$ to $0.2$, respectively) will need to be changed depending upon the amplitude and composition of the speech segment that you are working with, but if chosen correctly will ensure that the plots fill the entire plot area. This resulting plot is reproduced in Figure 2.9. It should be noted from the figure that the first major peak identified in the correlogram, at an $x$ distance of $0.006\,875$, which is $6.875$ ms, corresponds to the main period in the speech plot above. Try measuring the distance between speech peaks with a ruler, and then comparing this to the distance from the $y$-axis to the first identified peak in the correlogram. This illustrates the main use of the technique for audio analysis – detecting periodicities.

### 2.6.2.2 Cepstrum

The name 'cepstrum' comes about by reversing the first half of the word 'spectrum', and plots the amplitude of a signal against its 'quefrency' – actually the inverse frequency. Evidently neither word was chosen for ease of pronunciation. However, the technique is particularly good at separating the components of complex signals made up of several simultaneous but different elements combined together – such as speech (as we will see in Chapter 3).

The cepstrum is generated as the Fourier transform of the log of the Fourier transform of the original signal [4]. Yes, there really are two Fourier transform steps, although in practice the second one is often performed as an inverse Fourier transform instead [5].[2]

Using MATLAB again, a very simple example would be to plot the cepstrum of the speech analysed above with the correlogram. This is fairly simple to plot – not quite accurately as per the original meaning of cepstrum, but certainly useful enough:

```
ps=log(abs(fft(hamming(length(segment)).*segment)));
plot(abs(ifft( ps )));
```

Most likely, if the speech segment were as large as that used for the correlogram example, the resulting cepstrum would have a huge DC peak and much of the detail in the plot obscured. It is possible to zoom in on sections of the plot either using the MATLAB viewing tools, or by reducing the size of the original speech segment being analysed.

Within a cepstral plot, there should be a peak visible at the same index position as the peak in the correlogram. For example, a peak in the 256-point cepstrum of 8 kHz speech at $x$-index 56 would relate to a frequency of $4000 * 56/256 = 875$ Hz. This method of analysis will be illustrated later in Section 6.1.5.

It seems that both the correlogram and the cepstrum can reveal a fundamental frequency. Both methods, while accomplishing similar tasks, have unique strengths: peak

---

[2] Technically the method described here is the discrete time power cepstrum, arguably the most useful of the cepstral techniques.

detection in the cepstrum may be easier to achieve automatically (assume a constrained search range), and certainly the cepstrum can highlight features missed in a correlogram. The correlogram on the other hand is significantly easier to compute and scale than the cepstrum. Both are visualisation and analysis tools available to speech researchers, and many times both should be consulted during a particular analysis to compare their different viewpoints.

Finally, note that the cepstrum is often calculated with a nonlinear Mel frequency band mapping. The resulting Mel frequency cepstral coefficients (MFCC) are used for speaker identification, speech recognition and several other speech analysis tasks [6].

## 2.7    Sound generation

It is often quite useful to be able to generate sounds, especially when experimenting with, or testing, audio and hearing responses. Some of these techniques will come in useful during later chapters which provide MATLAB code to demonstrate several auditory phenomena.

### 2.7.1    Pure tone generation

A pure tone is an unmodulated sinusoid in shape. To generate such a tone of frequency Ft at a sample rate of Fs, lasting for Td seconds, is reasonably easy, and in fact we will create a small MATLAB function called tonegen() to do precisely this.

```
function [s]=tonegen(Ft, Fs, Td)
s=sin([1:Fs*Td]*2*pi*Ft/Fs);
```

Such a function, saved in the MATLAB work directory using a filename of tonegen.m, would then be accessible to other functions and programs written in MATLAB.

Thus, using this to generate and listen to a 2 second pure tone of frequency 440 Hz (corresponding to $A_4$ on the musical scale as described in Infobox 2.5 on page 33), we would do the following:

```
note=tonegen(440, 16000, 2);
soundsc(note, 16000);
```

### 2.7.2    White noise generation

White noise consists of uniformly distributed random numbers, which are easily generated in MATLAB, using the same notation as previously of Td seconds duration at sample rate Fs:

```
noise=rand(1,Fs*Td);
```

## 2.7.3 Variable tone generation

Occasionally it is necessary to generate a time-varying frequency. This is not quite as trivial as one may imagine, since when dealing with sinewaves, any discontinuity when changing frequency will be clearly audible. However the following MATLAB function will smoothly transition frequencies when fed with an array specifying desired frequency, sample-by-sample, in Hz:

```
function [snd]=freqgen(frc, Fs)
  th=0;
  fr=frc*2*pi/Fs;
  for si=1:length(fr)
    th=th+fr(si);
    snd(si)=sin(th);
    th=unwrap(th);
  end
```

We can test this function by creating an array specifying a few musical note frequencies, then convert this into sound and listen to the result:

```
freq=[440*(1+zeros(1,1000)), 415.2*(1+zeros(1,1000)),
        392*(1+zeros(1,1000))];
music=freqgen(freq, 8000);
soundsc(music, 8000);
```

## 2.7.4 Mixing sounds

Given two audio arrays of equal length we may combine or mix them together. We will illustrate this and listen to the result using soundsc() (since the result of the addition may well cause individual sample amplitudes to exceed the +/− 1.0 limit which would cause the sound() function to clip).

Here are notes of a chord generated and replayed in MATLAB:

```
C=tonegen(261.63, 8000, 2);
E=tonegen(329.63, 8000, 2);
G=tonegen(783.99, 8000, 2);
B=tonegen(987.77, 8000, 2);
soundsc(C+E+G+B, 8000);
```

For those readers with a musical theory background, it should be relatively easy to create MATLAB functions able to create specified chords, minors or majors and replay them on demand.

---

**Infobox 2.4** Visualisation of signals

---

As an example of the difficulty in analysing a constantly changing signal, we will deliberately construct a 'frequency agile' signal (one which changes rapidly in frequency characteristics) and then plot it:

```
y=chirp([0:0.001:5],0,5,500);    %construct a frequency chirp
z=[y,y[length(y):-1:1],y];       %make a zig-zag
```

First let's take one huge FFT and plot this as shown below:

```
f=abs(fft(z, 8192));
plot(f(1:4096));
```

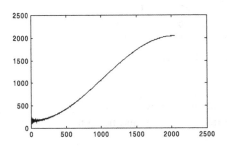

Absolute FFT plot of chirp spectrum.

Does this spectrum really represent the characteristics of the signal we have constructed? Let us explore further by slicing the signal up into smaller pieces and plotting as a waterfall plot to see how the frequency changes over time. We could do the slicing manually, but in this case we can be lazy and use spectrogram which accomplishes the same thing for us:

```
s=spectrogram(z, 1024);
waterfall(abs(s)'); % may take some time to process
```

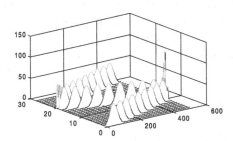

Absolute FFT plot for audio spectrum.

The waterfall plot above shows almost 30 slices in time, each one giving a 512-bin FFT plot, and clearly indicating that a different frequency component is present during each of the time periods. Evidently a totally different picture emerges of the signal in this plot compared to the spectrum plotted initially. If we had relied upon only an FFT to observe the signal characteristics we would have missed a significant amount of information.

Now explore the spectrogram function; without assigning its output to a variable, it plots the spectrogram directly on screen. Try changing the 1024 to 256 and then 8192 and see the difference it makes to the visualisation, and to your understanding of what the signal really is.

**Infobox 2.5** Musical notes

The pitch of a musical instrument note is primarily determined by its fundamental frequency of oscillation as perceived by a human. There may in actuality be many different frequencies present in that sound, generally harmonically related, and which through their various proportions affect the timbre of the note.

Many musical scales exist to define note frequencies, but the Western scale currently dominates, tied to a frequency reference where $A_4$ (the A above middle-C) is set to 440 Hz. Defined by the International Standards Organisation, this is often called *concert pitch*. Any reader who has played an instrument in an orchestra would know the importance of all instruments playing in tune. Before the 1940s, however, lack of standardisation meant that $A_4$ could vary widely between orchestras and countries, and many composers insisted on setting their own (incompatible) standards. Both Germany and the United Kingdom created early standards, but it was the British Broadcasting Corporation (BBC) who chose a 440 Hz standard based on the very high precision oscillation of a piezoelectric crystal acted upon by dividers and multipliers. The earlier 439 Hz standard, being a prime number, could not be generated using such methods.

Interested readers should note that the Western scale, chosen like others to match human perception of pitch differences, defines that the frequency relationship of one semitone is the twelfth root of two. Therefore if $A_4$ is 440 Hz, $A\sharp_4$ should be:

```
440*2^(1/12)
```

and since an octave consists of 12 semitones, $A_5$ would be:

```
440*(2^(1/12))^12=440*2=880
```

In this way, by knowing the semitone relationship between musical notes, we can determine the frequency of any note on the musical scale.

## 2.8     Summary

This chapter has covered the basics of obtaining and storing audio with MATLAB, some of the most important features to be aware of when working with audio, methods of handling and visualising audio and finally of creating sounds.

Subsequent chapters will build upon these techniques, incorporating them into more involved methods of analysis and processing. It would therefore be desirable to ensure familiarity with the techniques presented in this chapter before progressing into too much depth in subsequent chapters.

# Bibliography

Readers wishing to learn more about either using MATLAB itself, the underlying basics of signal processing, or the theories and practicalities of audio applications, may wish to consult some of the references given by category below:

## MATLAB

- *An Engineers Guide to MATLAB*
  E. B. Magrab, S. Azarm, B. Balachandran, J. Duncan, K. Herold, G. Walsh (Prentice-Hall, 3rd edition 2007)
  This is an expensive and extensive textbook running to 750 pages of engineering applications in MATLAB. Although it does not mention audio and speech processing explicitly, it does very nicely cover techniques of data handling, visualisation and analysis. File input and output are covered, as are statistical techniques. This may well be a useful reference work, consulted in the library, rather than purchased for continuous use.

- *A Guide to MATLAB: For Beginners and Experienced Users*
  B. Hunt, R. Lipsman, J. Rosenberg, K. R. Coombes, J. E. Osborn, G. J. Stuck (Cambridge University Press, 2nd edition 2006)
  This far more concise reference book dispenses with the applications focus and spends more effort in covering the basic and underlying techniques of data handling, plotting and so on. It is pitched at a lower level audience, but does claim a relevance for experienced users who are upgrading to a newer MATLAB version. The focus of this work is primarily the applied mathematics foundation of handling signals in MATLAB.

## Signal Processing

- *Digital Signal Processing: A Practical Guide for Engineers and Scientists*
  S. W. Smith (Newnes, 2002)
  Also freely available from **www.dspguide.com**
  See page 6 for further details.

- *Digital Signal Processing: A Practical Approach*
  E. C. Ifeachor and B. W. Jervis (Addison-Wesley, 1993)
  As the title implies, this work is reasonably practical in tone, and covers most of the required areas of signal processing, including some simple coverage of speech and audio. There are very many practical examples included and stepwise methods given where appropriate. Good end-of-chapter problems are provided. This book would be useful for brushing up on DSP, but may not suit an absolute beginner. Unfortunately it is marred by a handful of scattered mistakes.

- *Signal Processing First*
  J. McClellan, R. W. Schafer, M. A. Yoder (Pearson Education, 2003)
  A text for beginners, this book starts with introductory descriptions related to sound, and follows through to show how this can be represented digitally, or by computer. The coverage of basic Fourier analysis, sampling theory, digital filtering and discrete-time systems is gentle yet extensive. It is also possible to obtain a set of MATLAB examples related to the material in this book.

## Audio

- *A Digital Signal Processing Primer: With Applications to Digital Audio and Computer Music*
  K. Steiglitz (Prentice-Hall, 1996)
  There are few books that cover introductory audio and speech systems alone. This too covers audio as an application of digital signal processing rather than as a subject in its own right. The book thus spends time considering the DSP nature of handling audio signals, which is no bad thing; however the orientation is good, being practical and relatively focused on the applications.

- *Speech and Audio Signal Processing: Processing and Perception of Speech and Music*
  B. Gold and N. Morgan (Wiley, 1999)
  The epitome of speech and audio textbooks, this 560-page tome is divided into 36 chapters that cover literally every aspect of the processing and perception of speech and music. For readers wishing to purchase a single reference text, this would probably be first choice. It is not a book for absolute beginners, and is not orientated at providing practical methods and details, but for those already comfortable with the main techniques of computer processing of speech and audio, it would be useful in expanding their knowledge.

# References

[1] S. W. Smith. *Digital Signal Processing: A Practical Guide for Engineers and Scientists*. Newnes, 2000. URL **www.dspguide.com**.

[2] J. W. Gibbs. Fourier series. *Nature*, **59**: 606, 1899.

[3] R. W. Schaefer and L. R. Rabiner. Digital representation of speech signals. *Proc. IEEE*, **63**(4): 662–677, 1975.

[4] B. P. Bogert, M. J. R. Healy, and J. W Tukey. The quefrency analysis of time series for echoes: Cepstrum, pseudo-autocovariance, cross-cepstrum and saphe cracking. In M. Rosenblatt, editor, *Proceedings of the Symposium on Time-Series Analysis*, pages 209–243. John Wiley, 1963.

[5] D. G. Childers, D. P. Skinner, and R. C. Kemerait. The cepstrum: A guide to processing. *Proc. IEEE*, **65**(10): 1428–1443, 1977.

[6] F. Zheng, G. Zhang, and Z. Song. Comparison of different implementations of MFCC. *J. Computer Sci. and Technol.*, **16**(6): 582–589, 2001.

# 3 Speech

Chapter 2 described the general handling, processing and visualisation of audio vectors: sequences of samples captured at some particular sample rate, and which together represent sound. This chapter will build upon that foundation, and use it to begin to look at speech. There is nothing special about speech from an audio perspective – it is simply like any other sound – it's only when we hear it that our brains begin to interpret a particular signal as being speech. There is a famous experiment which demonstrates a sentence of sinewave speech. This presents a particular sound recording made from sinewaves. Initially, the brain of a listener does not consider this to be speech, and so the signal is unintelligible. However after the corresponding sentence is heard spoken aloud in a normal way, the listener's brain suddenly 'realises' that the signal is in fact speech, and from then on it becomes intelligible. After that the listener cannot 'unlearn' this fact: similar sentences which are generally completely unintelligible to others will be perfectly intelligible to this listener [1].

Apart from this interpretative behaviour of the human brain, there are audio characteristics within music and other sounds that are inherently speech-like in their spectral and temporal characteristics. However speech itself is a structured set of continuous sounds, by virtue of its production mechanism. Its characteristics are very well researched, and many specialised analysis, handling and processing methods have been developed over the years especially for this narrow class of audio signals.

Initially turning our back on the computer and speech processing, this chapter will consider the human speech production apparatus, mechanisms, and characteristics. This will be followed by an examination of the physical properties of speech itself resulting from the mechanism of its generation. We will then begin our study of how these properties allow for speech-related processing efforts.

## 3.1 Speech production

The sound that we know as speech begins with the lungs contracting to expel air, which carries sound of an approximately Gaussian frequency distribution [2]. This air is forced up through the bronchial tract past a set of muscle folds at the top of the trachea called vocal chords, and sets these vibrating. The air then enters the rear of the mouth cavity where it follows one of two paths to the outside. The first path is over and around the tongue, past the teeth and out through the mouth. The second route is through the nasal

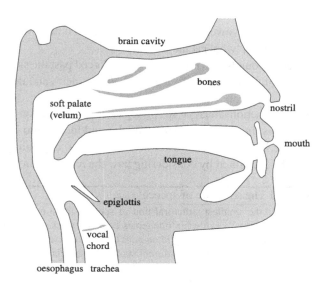

**Figure 3.1** Sectional diagram of human vocal apparatus, showing major articulators, resonators and features of the vocal and nasal tracts.

cavity – and this is the only route possible when the velum is closed. Figure 3.1 shows a diagram of the speech production apparatus (otherwise known as the human head).

The actual sound being produced depends on many criteria including the lung power and pressure modulation, the constriction at the glottis, the tension in the vocal chords, the shape of the mouth, and the position of the tongue and teeth. A brief overview follows, with more detail available in Infobox 3.3: Speech articulation, on page 44.

(a) Lung power mostly affects the volume of the sound, but rapid variation often distinguishes a boundary between syllables.
(b) If the glottis is closed temporarily during speech, a glottal stop results such as the /t/ in a Yorkshire-accented reading of 'I went t' shops'. A plosive sound like the /d/ in 'dog', is a short stop followed by an explosive release.
(c) Vocal chord muscle tension causes the chords to vibrate at different rates, forming the pitch frequencies. Voiceless sounds (e.g. /s/ in 'six'), where the vocal chords do not vibrate, have little or no pitch structure.
(d) If the air is diverted through the nose by the velum closing, a nasal sound such as /m/ in 'mad' results. Different timbre also results from the slightly different path length from lungs to nose compared with lungs to mouth (imagine two different length organ pipes).

**(e)** If the air travels through the mouth, a humped tongue and opening then closing lower jaw cause a vowel sound (e.g. /a/ in 'card'), if the lower jaw does not close, a glide (e.g. /w/ in 'won') is the result.

**(f)** Different sounds also result if the air is forced past the sides of a tongue touching the roof of the mouth or the teeth (e.g. /l/ in 'luck', and the /th/ sound).

The above actions must be strung together by the speaker in order to construct coherent sentences. In practice, sounds will slur and merge into one another to some extent, such as the latter part of a vowel sound changing depending on the following sound. This can be illustrated by considering how the /o/ sound in 'or' and in 'of' differ.

---

**Infobox 3.1** The structure of speech

A *phoneme* is the smallest structural unit of speech: there may be several of these comprising a single word. Usually we write phonemes between slashes to distinguish them, thus /t/ is the phoneme that ends the word 'cat'. Phonemes often comprise distinctly recognisable *phones* which may vary widely to account for different spoken pronunciations.

Two alternative pronunciations of a phoneme are usually the result of a choice between two phones that could be used within that phoneme. In such cases, the alternative phone pair are termed *allophones*. Interestingly, phones which are identical except in their spoken tone, can be called *allotones*, something which is very common in Mandarin Chinese, where many phonemes can be spoken with a choice of tone to totally change the meaning of a word.

Single or clustered phonemes form units of sound organisation called *syllables* which generally allow a natural rhythm in speaking. Syllables usually contain some form of *initial* sound, followed by a *nucleus* and then a *final*. Both the initial and the final are optional, and if present are typically consonants, while the syllable nucleus is usually a vowel.

Technically a *vowel* is a sound spoken with an open vocal tract as explained in Section 3.1, while a *consonant* is one spoken with a constricted, or partially constricted vocal tract, but as with many research areas, these definitions which are so clear and unambiguous on paper are blurred substantially in practice.

---

The merging of phonemes and words together is one major difficulty in speech processing – especially in the field of continuous speech recognition. For simple, single syllable words, the obvious gaps in a waveform plot will correspond to demarcation points, but as the complexity of an utterance increases, these demarcations become less and less obvious, and often the noticeable gaps are mid-word rather than between words. These difficulties have led speech segmentation to being a flourishing research area (see also Section 7.5.4).

Finally, spoken enunciation is context sensitive. When background noise is present we shout, during extreme quiet we whisper. This does not always hold true for communications channels: imagine a man in a quiet office telephoning his wife in a noisy shopping mall. The husband will naturally talk fairly quietly in order not to disturb his colleagues, but the wife will have to shout to be heard. Possibly the wife will ask the office-bound husband to speak up a little and the husband will then ask the wife to stop shouting – this example is related from personal experience.

## 3.2     Characteristics of speech

Despite many differences between individuals, and the existence of many languages, speech follows general patterns, and on average has well defined characteristics such as those of volume, frequency distribution, pitch rate and syllabic rate [3]. These characteristics have adapted with regard to environment, hearing and voice production limitations – speech characteristics fit the speech generating abilities of the body – but the rapid changes in society over the past century have exceeded our ability to adapt.

The shouting mechanism for 'long distance' communications, for example, across an open valley, is not particularly suited to inner-city conditions (just stand outside a tower block for a few minutes on a hot day when windows are open and hear the examples of inappropriately loud vocal communications). On the other hand, rail or bus commuters will seldom have the opportunity to converse in whispers.

---

**Infobox 3.2** The International Phonetic Alphabet

The International Phonetic Alphabet (IPA) is the usual method of describing and writing the various phonemes that make up speech. As defined by the International Phonetic Association, a set of symbols, written using a shorthand notation, describes the basic sound units of words. These symbols can completely describe many different languages using the 107 letters and several diacritical marks available [3]. It is beyond the scope of this book to introduce this alphabet, but simply to point out that researchers working with phonetics would be advised to learn the IPA and apply this notation in their work to avoid misconceptions and insufficiently specified speech sounds.

---

### 3.2.1     Speech classification

Physically, the sounds of speech can be described in terms of a pitch contour and formant frequencies. In fact this description forms a method of analysis used by most speech compression algorithms (discussed in Section 5.2 and beyond). Formants are resonant frequencies of the vocal tract which appear in the speech spectrum as clear peaks. As an example, three distinct formant peaks can be seen in the frequency domain plot of a short speech recording, in Figure 3.2.

Formants have been described by the famous researcher Klatt and others as being the single most important feature in speech communications [4]. Generally many formants will be present in a typical utterance, and the location of these will vary over time as the shape of the mouth changes. Formants are counted from the lowest frequency upwards, and usually only the first three (F1, F2 and F3) contribute significantly to the intelligibility of speech. Some fricative sounds like /ch/ can produce a lot of formants, but generally speaking F1 contains most of the speech energy while F2 and F3 between them contribute more to speech intelligibility [5].

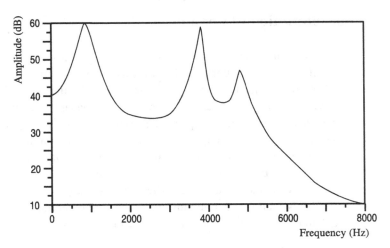

**Figure 3.2** Spectrum plot of a 20 ms recording of voiced speech, showing three distinct formant peaks.

The pitch contour (often called f0 – note the lower case notation) is the parameter that describes the tone of the voice (the perceived frequency), and is in effect the fundamental vocal frequency. Again, pitch frequencies contain energy but contribute little to intelligibility for English and other European languages [6]. It is, however, a very different matter in a tonal language such as Mandarin Chinese which is totally dependent on tone for conveying meaning [7,8]. As an example, in Chinese the single word 'ma' can mean one of five things depending on which tone it is spoken with: mother, horse, scold, question, etc. and this is not an isolated example since all single Chinese word sounds have multiple meanings differentiated by tone.

## 3.2.2     Amplitude distribution of speech

The overall amplitude distribution of speech depends upon the speaker's personality and mood (every reader is likely to have endured monotonous talks on occasion – literally meaning 'single tone' speech), environmental noise, infection, and so on.

Also feedback from a listener, either verbal 'speak up please' or non-verbal, such as cupping a hand around an ear, can prompt a speaker to alter their vocal characteristics. However despite this variability, it is interesting to determine average speech levels in different environments, as shown in Table 3.1, reproduced from [9], where the sound amplitudes are listed in $dB_{SPL}$.[1]

Note the wide range in speech level, and the relationship between that and the location. In a train and in an aeroplane are the only situations listed where a negative

---

[1] dB (decibel) is a base 10 logarithmic measure of amplitude, with the *SPL* subscript in $dB_{SPL}$ referring to sound pressure level, which is referenced so that 0 dB is the quietest average audible sound at 1 kHz. In terms of measurable pressure, $74\,dB_{SPL}$ is 1 $\mu$bar, 1 dyne $cm^{-2}$ or 0.1 Pa in different units.

**Table 3.1.** Amplitude of speech in several environments, from [9].[a]

| Location | Noise level (dB$_{SPL}$) | Speech level (dB$_{SPL}$) |
|---|---|---|
| school | 50 | 71 |
| home (outside, urban) | 61 | 65 |
| home (outside, suburban) | 48 | 55 |
| home (inside, urban) | 48 | 57 |
| home (inside, suburban) | 41 | 55 |
| department store | 54 | 58 |
| on a train | 74 | 66 |
| in an aircraft | 79 | 68 |

[a] These data were originally published in *The Handbook of Hearing and the Effects of Noise*, K. Kryter, Chapter 1, Copyright Elsevier (Academic Press) 1994.

signal-to-noise level results (i.e. where the speech level is lower than that of the noise), and that is partly due to the particular shape of the noise spectrum in both environments. Later, in Chapter 6 we will derive a way of taking account of differences in speech and noise spectra.

In general, as the noise level increases by 1 dB, a speaker will raise his voice level by 0.5 dB within the range of normal speech [9]. With very low noise levels, a male adult speaker can produce 52 dB$_{SPL}$ of speech measured at a distance of 1 m when speaking casually. This raises to about 90 dB$_{SPL}$ when shouting. Quoted figures of 50–82 dB$_{SPL}$ for women and 53–83 dB$_{SPL}$ for children were probably obtained by someone with no experience of child raising.

The dynamic range of conversational speech is around 30 dB [10], and the mean level for males measured at 1 m is somewhere in the region of 55–60 dBA$_{SPL}$[2] with peaks that extend 12 dB beyond this [6].

## 3.2.3    Types of speech

Speech can be classified into phonemes which fall into several different types (see Infobox 3.1: The structure of speech on page 40), whereas the phonemes themselves are best identified using the international phonetic alphabet (see Infobox 3.2: The International Phonetic Alphabet on page 41).

When spoken, phonemes will naturally have different average amplitudes. These are shown classified by type in Table 3.2, where the relative amplitude of spoken phonemes in each class is listed, along with the amplitude range of phonemes within that class. More detail, including phoneme-by-phoneme examples from the original experiments, can be found in [11].

---

[2] The 'A' in dBA refers to the A-weighting curve (discussed in Section 4.2.1), in which a frequency correction is made to the measured signal prior to obtaining the average amplitude. This frequency correction is based upon 'hearing curves' and attempts to make the measured value more representative of what a human would perceive.

**Table 3.2.** Average amplitude of phonemes by class, also showing amplitude range within each class, measured with respect to the quietest phoneme in English, the voiceless fricative /th/ in 'thought'.

| Phoneme class | Example | Amplitude (range), dB |
|---|---|---|
| vowel | card | 26.0 (4.9) |
| glide | luck | 21.6 (3.2) |
| nasal | night | 17.1 (3.0) |
| affricative | jack | 14.9 (2.6) |
| voiced fricative | azure | 11.5 (2.2) |
| voiceless fricative | ship | 10.0 (10.0) |
| voiced plosive | bap | 9.6 (3.3) |
| voiceless plosive | kick | 9.5 (3.3) |

Without exception, vowels are spoken with more power than other phonemes, and the range of intensity for all listed sounds is rather large, in fact it spans almost 30 dB, but remember that these figures have necessarily been time averaged – the instantaneous differences may be even higher.

A useful rule of thumb is that, in normal speech, vowels are approximately 12 dB louder than consonants. This is perhaps surprising given that, in English at least, consonants convey a greater share of vocal intelligibility than vowels. I generally ask sceptics who require a demonstration of the relative information carrying content of vowels with respect to consonants to read aloud the following sentence:

*The yellow dog had fleas.*

Next replace all consonants with the same phoneme and read again:

*Tte tettot tot tat tteat.*

---

**Infobox 3.3** Speech articulation

Many sounds, both consonant and vowel, are defined by their place, or method of articulation within the vocal tract. Here is a list of some of the more common terms:

- *affricative* – a turbulent airflow fricative following an initial stop. E.g. /ch/ in 'chip'.
- *diphthong* – a two-part sound consisting of a vowel followed by a glide. E.g. /i//n/ in 'fine'.
- *fricative* – a very turbulent airflow due to a near closure of the vocal tract. E.g. /sh/ in 'ship'.
- *glide* – a vowel-like consonant spoken with almost unconstricted vocal tract. E.g. /y/ in 'yacht'.
- *nasal* – a consonant spoken with vellum lowered, so sound comes through the nasal cavity. E.g. /m/ in 'man'.
- *stop* or *plosive* – an explosive release of air upon rapid removal of a vocal tract closure. E.g. /p/ in 'pop'.

Most of the consonant sounds can be either voiced or unvoiced, depending upon whether the glottis is resonating. For example /c/ in 'cap' is unvoiced whereas /g/ in 'gap' is voiced.

Finally, replace all vowels with the same phoneme and read again:

*Tha yallaw dag had flaas.*

Apart from utterly humiliating such sceptics by making them sound stupid, it is immediately obvious that although the same-vowel sentence sounds odd, it is still highly intelligible. By contrast the same-consonant sentence is utterly unintelligible. This simple example illustrates that although vowels are spoken louder, they tend to convey less intelligibility than the quieter consonants.

### 3.2.4    Frequency distribution

The frequency distribution of speech follows fairly closely to the sensitivity of the human ear: most of the frequencies involved in speech, and certainly all of those that convey significant intelligibility, lie within the range of frequencies over which the ear is most sensitive. However within this band (about 300 Hz to 4 kHz) there is a mismatch between the speech frequencies of greatest energy, and those of greatest intelligibility. Put another way, the speech frequencies with the greatest concentration of power are not quite the same as those that account for most transmitted intelligibility [12] – this disparity is hinted at by the vowel/consonant difference in the previous section. To examine further, let us now consider both power and intelligibility separately.

**Speech power:** Most of the energy transmitted in speech is based at low frequencies, approximately 500 Hz for males and 800 Hz for females. These frequencies are not essential for intelligibility – experiments in which these frequencies are removed indicate that the remaining speech, whilst quiet and unusual in sound, can still be perfectly intelligible. That is to say, the spoken information in the speech remains, whereas by contrast the ability to recognise the speaker is severely impaired. Typically around 84% of the energy in speech is located below 1 kHz as shown in Figure 3.3 constructed from data provided in [6] and [13].

Figure 3.3 also contains bands indicating the ranges where the first three formants usually lie. Note the correspondence between F1 and the band of greatest energy distribution.

**Intelligibility:** Although this will be discussed more fully in the next section, it is worth noting here that most of the information (in terms of what was said) transmitted by speech lies above 1 kHz, carried by formants F2 and F3 as mentioned previously. Removal of all speech energy between 800 Hz and 3 kHz would leave a signal that sounded like speech but which was completely unintelligible [14]. This effect is illustrated in Figure 3.4.

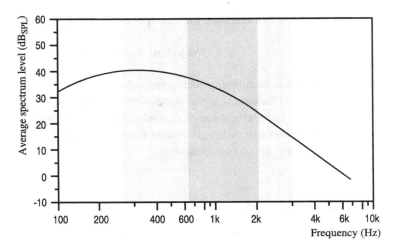

**Figure 3.3** Long-time averaged speech power distribution plotted against frequency, with the approximate regions of the first three formants identified through vertical grey bands.

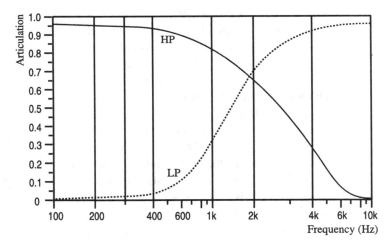

**Figure 3.4** Effect of limiting speech frequency range on the intelligibility of speech syllables, measured as articulation index.

An analysis of the figure reveals that if a speech signal were low-pass filtered at 1 kHz, around 25% of speech syllables would be recognisable. If it were high-pass filtered at 2 kHz, around 70% would be recognisable.

### 3.2.5      Temporal distribution

Temporally-wise, the major constraint on speech is how fast the brain and vocal apparatus can attempt to articulate phonemes or syllables. The various muscles involved in vocal production can only move so fast, as can the muscles controlling the lungs. A further constraint on lung muscle movement is the need for regular lung re-filling required to prevent asphyxia.

Evidence suggests that the speed of articulation is mostly independent of the rate of speaking. Even when speaking more quickly, most people will use the same length of time to articulate a particular syllable, but will reduce the length of the gaps between syllables and words [4]. This uniformity does, of course, greatly assist in the artificial description and modelling of speech.

Of all the constraints on speech, the speed at which the muscles are capable of moving is the most useful to us [15]. It allows speech to be defined as semi-stationary over periods of about 20 ms, meaning that speech analysis (including short-time Fourier analysis, linear predictive analysis and pitch detection) is usually conducted over such a duration. This is termed pseudo-stationarity (see Section 2.5.1).

One further temporal generalisation is that of syllabic rate, the rate at which syllables are articulated. For most languages this remains fairly constant, and varies only slightly between individuals and types of speech [6].

One or more agglomerated phonemes can make up a syllable sound. For humans, it seems that the simplest unit of recognition may be the syllable, whereas the phoneme is a distinction generally made by speech researchers aiming to determine a set of basic building blocks for speech sounds. Phoneme duration varies from language-to-language, speaker-to-speaker and differs depending upon the exact phoneme. There is even evidence to tie phoneme length to word stress – louder, and more emphasised phonemes tend to exhibit a longer duration.

## 3.3 Speech understanding

Up to now, this chapter has investigated the production of speech and the characteristics of the produced speech. Ignoring aspects of whichever communications mechanism has been used, this section will now consider some of the non-auditory factors involved in the understanding of speech by humans. That is, the nature of speech structure and how that relates to understanding, rather than the nature of human hearing and perception of speech (which will be covered in Chapter 4).

### 3.3.1 Intelligibility and quality

Firstly it is very important to distinguish between these two terms. Both are correctly used interchangeably at times, but their measurement and dependencies are actually very different. In very simple terms, quality is a measure of the fidelity of speech. This includes how well the speech under examination resembles some original speech, but extends beyond that to how nice the speech sounds. It is a highly subjective measure, but can be approximated objectively.

**Table 3.3.** Average amplitude of several spoken speech types.

| Score | Description | Impairment |
|-------|-------------|------------|
| 5 | excellent | imperceptible |
| 4 | good | perceptible but not annoying |
| 3 | fair | slightly annoying |
| 2 | poor | annoying |
| 1 | bad | very annoying |

Intelligibility is a measure of how understandable the speech is. In other words, it concentrates on the information-carrying content of speech. Some examples should clarify the difference:

1. A string of nonsense syllables, similar to baby speech, spoken by someone with a good speaking voice can sound very pleasant, of extremely high quality, but contains no verbal information, in fact has no intelligibility at all.
2. A recording of speech with a high-frequency buzzing sound in the background will be rated as having low quality even though the words themselves may be perfectly understandable. In this case the intelligibility is high.
3. When choosing a car audio system, you might tune to a favourite radio station to test. Generally the audio system that sounds nicest (of highest quality) would be the one purchased.
4. When the military are in a combat situation, it is usually extremely important to understand the speech from a radio, whereas the quality of the sound is almost totally unimportant. In World War II, crude speech processing (clipping, or filter-clipping) was applied to radios used in aircraft – making the speech sound shrill and screechy – but significantly improving its intelligibility in a noisy cockpit [5]. This effect can often be heard in films and documentaries of the period.

Despite stressing the difference between quality and intelligibility in this section, it is useful to note that under most circumstances excellent intelligibility implies excellent quality, and very poor intelligibility implies very poor quality. These are the extremes – between these points the relationship between the two is not straightforward.

## 3.3.2    Measurement of speech quality

Speech quality is normally measured subjectively, in terms of a mean opinion score (MOS). This involves a panel of several listeners, usually placed in a soundproofed room, having the audio recording under evaluation played to them. They will then rate this according to the scale shown in Table 3.3.

The MOS score of a particular recording is the mean of the results reported by each of the listeners. Obviously the more listeners, the more accurate (and repeatable) the results will be. This test is standardised by the International Telecommunications Union (ITU) in recommendation P.800, widely used in the audio community.

In recent years, several objective quality algorithms have been developed. These are computer programs that, when fed an original audio recording, plus a degraded one, will estimate the MOS score of the degraded audio (or will produce a figure which can be converted to a MOS score). These programs can be tricked, and at the extremes will track actual MOS scores very poorly, but for normal speech processed in standard ways, have been shown to produce respectable results. They are used primarily because they are cheaper and quicker than forming a panel of listeners. Most importantly they allow automated testing to take place. Some of the more prominent algorithms are:

- PESQ (perceptual evaluation of speech quality);
- PSQM (perceptual speech quality measure);
- MNB (measuring normalised blocks).

Although these are commercially supported algorithms, it has been possible in the past to download working versions from the Internet for non-commercial research use.

A far more crude measure of quality between a processed audio vector $p$ and an original audio vector $s$ is the mean-squared error (MSE) $E$. This is simply calculated on a sample-by-sample basis as the average squared difference between those vectors:

$$E = \frac{1}{N} \sum_{i=0}^{N-1} \{s[i] - p[i]\}^2. \tag{3.1}$$

In MATLAB, without using library functions, that would be:

```
mse=mean((s-p).^2)
```

For long recordings, this measure would smear together all features of speech which change over time into one average. It is often more useful to know the mean-squared error on a segment-by-segment basis to see how this varies (think of a typical speech system you are developing: it would be more useful to know that it works very well for voiced speech but not well for unvoiced speech rather than know the overall average). The reader may remember the same argument used previously in Section 2.4 for a similar analysis example.

The segmental mean-squared error is a measure of the time-varying MSE. Usually segments are 20–30 ms in size, and are sequential analysis frames, sometimes with overlap. For a frame size of $N$ samples and no overlap, for the $j$th segment this would be:

$$E(j) = \frac{1}{N} \sum_{i=jN}^{(j+1)N-1} \{s[i] - p[i]\}^2. \tag{3.2}$$

```
mse(j)=mean((s(j*N+1:(j+1)*N)-p(j*N+1:(j+1)*N)).^2);
```

Remember that MATLAB indexes arrays from element 1 and not element 0, hence the slight difference to indexing terms between the given equation and the MATLAB expression.

For cases when signals of interest are not really being compared for likeness, but rather one signal is corrupted by another one, the ratios of the signals themselves can be useful, as the signal-to-noise ratio (SNR). Note that SNR is not used to measure the degree of difference between signals, but is simply the base-10 logarithm of the ratio between a wanted signal $s$ and interfering noise vector $n$, measured in decibels, dB.

$$\text{SNR} = 20 \log_{10} \left\{ \frac{1}{N} \sum_{i=0}^{N-1} \left( \frac{s}{n} \right) \right\}. \tag{3.3}$$

```
snr=10*log10 (s./n)
```

Segmental signal-to-noise ratio (SEGSNR) is a measure of the signal-to-noise ratio of segments of speech, in the same way that we segment other measures to see how things change over time. Again, segments are typically 20–30 ms in size, perhaps with some overlap. For a frame size of $N$ samples and no overlap, for the $j$th segment this would be as shown in Equation (3.4):

$$\text{SEGSNR}(j) = 20 \log_{10} \left\{ \frac{1}{N} \sum_{i=jN}^{(j+1)N-1} \left( \frac{s}{n} \right) \right\}. \tag{3.4}$$

```
segsnr(j)=10*log10(s(j*N+1:(j+1)*N)./n(j*N+1:(j+1)*N))
```

However on the basis that hearing is not conducted in the time domain, but in the frequency domain, such measures are likely to be only minimally perceptually relevant. Better measures – or at least those more related to a real-world subjective analysis – can be obtained in the frequency domain. The primary measure of frequency-domain difference is called spectral distortion, shown in Equation (3.7), measured in dB$^2$, as a comparison between signals $p(t)$ and $s(t)$ for a frame of size $N$.

First, however, since the equation is in the frequency domain, we convert the time-domain signals to be compared, into the frequency domain using the Fourier transform:

$$S(\omega) = \frac{1}{\sqrt{2\pi}} \int_{-\infty}^{+\infty} s(t) e^{-j\omega t} dt \tag{3.5}$$

$$P(\omega) = \frac{1}{\sqrt{2\pi}} \int_{-\infty}^{+\infty} p(t) e^{-j\omega t} dt \tag{3.6}$$

$$\text{SD} = \frac{1}{\pi} \int_{0}^{\pi} (\log(S(\omega)) - \log(P(\omega)))^2 d\omega. \tag{3.7}$$

In terms of implementation, and performing such analysis in MATLAB, we note firstly that performing an analysis from $-\infty$ to $+\infty$ is an unrealistic expectation, so we would normally choose a segment of $N$ samples of audio to analyse over, then window it and perform a fast Fourier transform to obtain both power spectra, $P$ and $S$. In discrete sampled versions, we follow the same method that we used in Chapter 2 to visualise signals in the frequency domain:

```
S=fft(s.*hamming(N));
S=20*log10(abs(S(1:N/2)));
P=fft(p.*hamming(N));
P=20*log10(abs(P(1:N/2)));
```

and then proceed with the SD measure:

```
SD=mean((S-P).^2);
```

Indeed, SD is a perceptually relevant difference measure for speech and audio, however it *can* be enhanced further, and that is by the additional step of A-weighting the spectra – so that differences in frequency regions that are more audible are weighted more than those in frequency regions that are inaudible. This yields a perceptually-weighted spectral distortion, and is used in practical systems that perform high-quality speech and audio signal analysis.

### 3.3.3    Measurement of speech intelligibility

Intelligibility is also best measured by a panel of listeners, and relates to the ability of listeners to correctly identify words, phrases or sentences. An articulation test is similar, but applies to the understanding of individual phonemes (vowels or consonants) in monosyllabic or polysyllabic real or artificial words. Several common methods of evaluation exist but those standardised by ANSI (in standard S2.3-1989) dominate. Some example evaluative procedures are listed here along with references that provide more information (unless noted, see [16] for further details):

- diagnostic rhyme test (DRT) [17] – asking listeners to distinguish between two words rhyming by initial, such as {freak, leak};
- modified rhyme test (MRT) – asking listeners to select one of six words, half differing by initial and half by final, such as {cap, tap, rap, cat, tan, rat};
- phonetically balanced word lists – presenting listeners with 50 sentences of 20 words each, and asking them to write down the words they hear;
- diagnostic medial consonant test;
- diagnostic alliteration test;
- ICAO spelling alphabet test;
- two-alternative forced choice [18] – a general test category that includes the DRT;
- six-alternative rhyme test [18] – a general test category that includes the MRT;

- four-alternative auditory feature test [17] – asking listeners to select one of four words, chosen to highlight the intelligibility of the given auditory feature;
- consonant-vowel-consonant test [19,20,11] – test of vowel syllable sandwiched between two identical consonants, with the recognition of the vowel being the listeners' task. For example {tAt}, {bOb};
- general sentence test [11] – similar to the phonetically balanced word list test, but using self-selected sentences that may be more realistic in content (and in context of what the test is trying to determine);
- general word test [5] – asking listeners to write down each of a set of (usually 100) spoken words, possibly containing realistic words.

Clearly intelligibility may be tested in terms of phonemes, syllables, words, phrases, sentences, paragraph meaning, and any other arbitrarily grouped, measured recognition rate. In general we can say that the smaller the unit tested, the more able we are to relate the results to individual parts of speech. Unfortunately no reliable method has so far been developed of extrapolating from, for example the results of a phoneme test, to determine the effectiveness on sentence recognition (although if you know what the cause of intelligibility loss is in a particular system, you could have a good guess).

## 3.3.4    Contextual information, redundancy and vocabulary size

Everyday experience indicates that contextual information plays an important role in the understanding of speech, often compensating for an extreme lack of original information. For example the sentence:

> *'He likes to xxxxx brandy'*

can easily be understood through guessing even though a complete word is missing ('drink').

The construction of sentences is such that the importance of missing words is very difficult to predict. It is hard to know in advance whether the start, middle or end of a sentence will be more critical to its understanding. For example the missing word 'stop' differs in both importance and predictability in the two sentences:

> *'She waited in the long queue at the bus xxxx'*

and

> *'As the car sped towards him he shouted xxxx!'*

Contextual information may be regarded as being provided by surrounding words which constrain the choice of the enclosed word, or on a smaller scale, by the surrounding syllables which constrain the choice of a missing or obscured syllable (as certain combinations do not appear at all, or very infrequently in the English language). Vocabulary size reduction also causes a similar constraint, and it is noticeable that most people will restrict their vocabulary to simple words when communications are impaired: eloquence is uncommon in highly noisy environments.

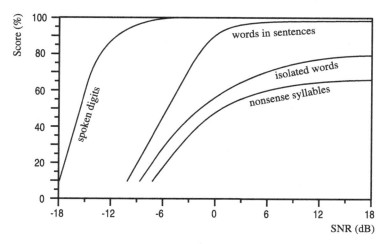

**Figure 3.5** Effect of contextual information on the intelligibility of speech.

Contextual information is also provided in other ways which may be situational, or may be through shared knowledge between two communicating parties. Redundancy, in which information is imparted in more ways than would normally be necessary, has effects similar to contextual constraint. Redundancy may be provided in the form of over-complex sentences in which the role of a single word is limited, by context, to a very small set of choices. In this way, the extra information given in the sentence should enable complete understanding even if some words were lost. A simple example would be the repetition of important words such as 'It is a green car. The green car is now turning.'

Redundancy may also be achieved by the use of longer phrases of words or descriptions, such as the use of 'alpha bravo foxtrot' instead of spelling out 'ABF' (these *anti-abbreviations* are from the NATO phonetic alphabet). Such redundancy reduces the chance that a short interruption to the audio (perhaps interference during transmission) would obliterate the entire meaning of the word.

A measurement of the effects of contextual information on understanding is extremely difficult to quantify and is, of course, highly subjective. It is also entirely dependent upon the testing method, and the method of information removal necessary to conduct the tests.

However some comparisons may be useful. Figure 3.5 attempts to illustrate this by plotting the percentage of correctly identified digits, syllables or words spoken in the presence of the given degree of background noise [11]. Although the shorter digit words are relatively more likely to be corrupted by noise than the predominantly longer unconstrained words, the extremely limited set of possible choices involved (as the listeners knew they were only listening for digits) means that even a corrupted digit may be guessed with some accuracy.

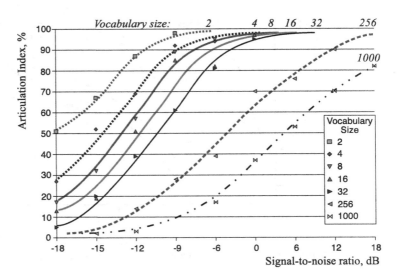

**Figure 3.6** Effect of spoken vocabulary size on the intelligibility of speech.

Vocabulary size can similarly be explored as shown next in Figure 3.6, where listeners are given the number of choices as indicated in the graph (i.e. the vocabulary size), for identification of spoken words. This illustrates the large improvement in recognition when vocabulary size is constrained whether artificially, or by context. For example, reducing vocabulary size from 256 to 16 at −9 dB signal-to-noise level, results in almost four times as many words being recognised. It should be noted that the articulation index shown is a measure of the recognition rate of individual phonemes, and not words themselves [10]. The figure was plotted from curve-fitting to tabular data presented in [11], derived from original experimental results in [21].

## 3.4    Summary

In this chapter we have studied the physical fundamentals of the speech signal – its amplitude, frequency distribution, tempo, and so on. We then considered the units of speech themselves: phonemes, words and sentences, and how these convey information.

We have looked at vocabulary effects and understanding, and touched on the major topics of quality and intelligibility and the testing of each of these.

We will now leave the subject of speech for a while. In Chapter 4 we move on to the ear and human hearing. Chapter 5 will then tie the hearing topic back into the characteristics of speech to develop the subject of perceived speech within communications systems.

## Bibliography

- *The Noise Handbook*
  Ed. W. Tempest (Academic Press, 1985)

- *Digital Processing of Speech Signals*
  L. R. Rabiner and R. W. Schafer (Prentice-Hall, 1978)

- *Computer Speech Processing*
  F. Fallside and W. Woods (Prentice-Hall, 1985)

- *Acoustic and Auditory Phonetics*
  K. Johnson (Blackwell Publishing, 2003)

- *The Handbook of Hearing and the Effects of Noise*
  K. Kryter (Academic Press, 1994)

- *The Physics of Speech*
  D. B. Fry (Cambridge University Press, 1979)

- *Speech Intelligibility and Speaker Recognition*
  Ed. M. E. Hawley (Halsted Press/Dowden Hutchinson and Ross, 1977)

# References

[1]   J. Barkera and M. Cooke. Is the sine-wave speech cocktail party worth attending? *Speech Comm.*, **27**(3-4): 159–174, 1999.

[2]   M. R. Schroeder, B.S. Atal, and J.L. Hall. Optimizing digital speech coders by exploiting masking properties of the human ear. *J. Acoustical Soc. America*, **66**(6): 1647–1652, 1979.

[3]   I. Witten. *Principles of Computer Speech*. Academic Press, 1982.

[4]   B. C. J. Moore. *An Introduction to the Psychology of Hearing*. Academic Press, 1992.

[5]   I. B. Thomas. The influence of first and second formants on the intelligibility of clipped speech. *J. Acoustical Soc. America*, **16**: 182–185, 1968.

[6]   J. Pickett. *The Sounds of Speech Communication*. Allyn and Bacon, 1980.

[7]   Z. Li, E. C. Tan, I. McLoughlin, and T. T. Teo. Proposal of standards for intelligibility tests of Chinese speech. *IEE Proc. Vision Image Sig. Proc.*, **147**(3): 254–260, 2000.

[8]   F. L. Chong, I. McLoughlin, and K. Pawlikowski. A methodology for improving PESQ accuracy for Chinese speech. In *Proc. IEEE TENCON*, Melbourne, November 2005.

[9]   K. Kryter. *The Handbook of Hearing and the Effects of Noise*. Academic Press, 1994.

[10]  L. L. Beranek. The design of speech communications systems. *Proc. IRE*, pages 880–890, 1947.

[11]  W. Tempest, editor. *The Noise Handbook*. Academic Press, 1985.

[12]  M. Mourjopoulos, J. Tsoukalas, and D. Paraskevas. Speech enhancement using psychoacoustic criteria. In *Proc. Int. Conf. on Acoustics, Speech and Signal Processing*, pages 359–362, 1991.

[13]  F. White. *Our Acoustic Environment*. John Wiley, 1976.

[14]  P. J. Blamey, R. C. Dowell, and G. M. Clark. Acoustic parameters measured by a formant estimating speech processor for a multiple-channel cochlear implant. *J. Acoustical Soc. America*, **82**(1): 38–47, 1987.

[15]  R. W. Schaefer and L. R. Rabiner. Digital representation of speech signals. *Proc. IEEE*, **63**(4): 662–677, 1975.

[16]  A. D. Sharpley. Dynastat webpages, 1996–2006. URL http://www.dynastat.com/Speech Intelligibility.htm.

[17]  S. F. Boll. Suppression of acoustic noise in speech using spectral subtraction. *IEEE Trans. Acoustics, Speech Signal Proc.*, **27**(2): 113–120, 1979.

[18]  R. E. P. Dowling and L. F. Turner. Modelling the detectability of changes in auditory signals. *Proc. Int. Conf. on Acoustics, Speech and Signal Processing*, Vol. 1 pages 133–136, 1993.

[19]  J. I. Alcantara, G. J. Dooley, P. J. Blamey, and P. M. Seligman. Preliminary evaluation of a formant enhancement algorithm on the perception of speech in noise for normally hearing listeners. *J. Audiology*, **33**(1): 15–24, 1994.

[20]  J. G. van Velden and G. F. Smoorenburg. Vowel recognition in noise for male, female and child voices. *Proc. Int. Conf. on Acoustics, Speech and Signal Processing*, pages 796–799, 1995.

[21]  G. A. Miller, G. A. Heise, and W. Lichten. The intelligibility of speech as a function of the context of the test materials. *Experi. Psychol.*, **41**: 329–335, 1951.

# 4    Hearing

A study of human hearing and the biomechanical processes involved in hearing, reveals several nonlinear steps, or stages, in the perception of sound. Each of these stages contributes to the eventual unequal distribution of subjective features against purely physical ones in human hearing.

Put simply, what we *think* we hear is quite significantly different from the physical sounds that may be present (which in turn differs from what would be captured electronically by, for example, a computer). By taking into account the various nonlinearities in the hearing process, and some of the basic physical characteristics of the ear, nervous system, and brain, it is possible to account for the discrepancy.

Over the years, science and technology has incrementally improved the ability to model the hearing process from purely physical data. One simple example is that of A-law compression (or the similar $\mu$-law used in some regions of the world), where approximately logarithmic amplitude quantisation replaces the linear quantisation of PCM: humans tend to perceive amplitude logarithmically rather than linearly, and thus A-law quantisation using 8 bits sounds better than linear PCM quantisation using 8 bits. It thus achieves a higher degree of subjective speech quality than PCM for a given bitrate [1].

## 4.1    Physical processes

The ear, as shown diagrammatically in Figure 4.1, includes the pinna which filters sound and focuses it into the external auditory canal. Sound then acts upon the eardrum where it is transmitted and amplified by the three bones, the malleus, incus and stapes, to the oval window, opening on to the cochlea. The figure shows the cochlea shaded in grey and drawn straight for clarity, whereas in reality it is somewhat coiled.

The cochlea, as a coiled tube, contains an approximately 35 mm long semi-rigid pair of membranes (basilar and Reissner's) enclosed in a fluid called endolymph [2]. The basilar membrane carries the organs of Corti, each of which contains a number of hair cells arranged in two rows (approximately 3500 inner and 20 000 outer hair cells).

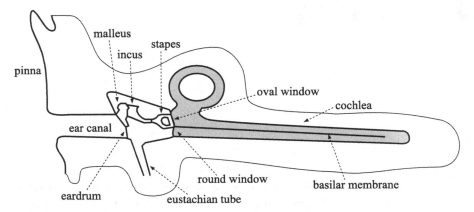

**Figure 4.1** Schematic and stylised cutaway diagram of the human ear, with cochlea unrolled and straightened to more clearly highlight its gently tapering shape.

Small electrical impulses are generated by the hair cells when tension is applied to them (usually when they are pulled, so that auditory response occurs during the rarefaction part of an oscillation in the fluid surrounding them). The width and stiffness of the basilar membrane tapers along its length, and thus when a vibration is applied to the system, the location on the basilar membrane which resonates is dependent upon frequency [3]. This process provides some frequency selectivity to the ear, with active processing by the auditory cortex providing the rest. In fact the auditory cortex processing substantially improves frequency selectivity. Nerves carry electrical impulses from the ear into the brain where extensive unconscious processing occurs [4]. The nature of the processing is not fully understood; however it is highly nonlinear, involves correlation between signals from each ear, and possibly with other senses (such as the feel of vibrations, and sight). The ear may be assumed to integrate over short times and also process more grossly repetitive patterns in different ways [4].

Active mechanisms also operate in the ear, which include tensioning the muscles operating on the malleus and stapes, to protect the ear from loud sounds, reducing the firing rate of certain hair cells and reportedly even producing sounds on occasions [4,5]. Further evidence for this is that audio acuity is reduced by illness or the taking of certain drugs, and earache can be caused by the failure of the inner ear muscles to protect the eardrum [6].

## 4.2    Psychoacoustics

An abbreviated concatenation of 'the psychology of acoustics', psychoacoustics describes the discrepancy between a purely physical view of hearing, and a subjective view. It is beyond the scope of this book to delve into too much detail concerning the psychological reasons behind this (and indeed it is likely that many of those reasons have not yet been scientifically identified). However extensive experiments have been

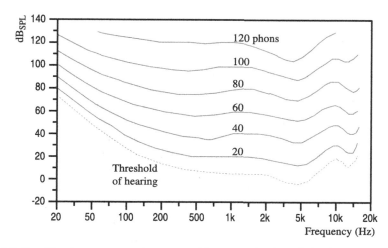

**Figure 4.2** Equal loudness contours (constructed from examination of Fletcher–Munson data).

conducted by many researchers over the many years, which allow a model of the effects of psychoacoustics to be built up. In the subsections that follow, several interesting phenomena are described that have a direct impact on the subject, and which will form the foundations of many of the techniques described in Chapter 7.

## 4.2.1 Equal loudness contours

Human subjects do not always judge tones of different frequency, but identical amplitude, to be equal in loudness [5,7]. Put another way, a 1 kHz signal at 40 dB$_{SPL}$ would sound louder than a 15 kHz signal at the same amplitude. Actually it is not surprising since 15 kHz is near the limit of hearing.

After long and careful research, scientists identified 'equal loudness contours' in the 1940s that model this behaviour. These contours are plotted in Figure 4.2, and show the curves identified in phons, where a curve of $n$ phons follows the contour of different frequency amplitudes that sound equally loud to a subject as an $n$ dB$_{SPL}$ tone at 1 kHz. The lowest curve (the dashed line) indicates the threshold of hearing.

The effect of equal loudness should be relatively easy to test in MATLAB, but one should beware that dependent upon the frequency response of the audio loudspeakers or headphones being used, some sounds will naturally be reproduced with higher amplitude than others. Assuming a good pair of headphones with relatively flat frequency response is available, let us create three pure tones in MATLAB using the tonegen() function from Section 2.7.1, and listen to them:

```
lo= tonegen(250, 441000, 2);
mi= tonegen(1200, 441000, 2);
```

```
hi= tonegen(11000, 441000, 2);
soundsc(lo, 441000);
soundsc(mi, 441000);
soundsc(hi, 441000);
```

Based on the loudness curves, the 1.2 kHz middle tone should be perceived as being louder than either the 250 Hz lower tone or the very high 11 kHz sound (which may not even be audible at all for many people). We have used a 44.1 kHz sample rate in this case in order to faithfully reproduce the 11 kHz sinewave.

It should be noted that the equal loudness contours derive from averaged tests, and apply to single tones only: there may be substantial differences from person-to-person, especially in cases of people who listened to too much rock music when young (suffering from hearing damage), but in the experience of the author, the curves do describe the ensemble of human responses fairly accurately.

---

**Infobox 4.1** Weighting

---

For speech and hearing purposes, voice power, background noise and other sound levels are usually measured in dBA, where the signal is A-weighted before being quantified. This is the application of a frequency weighting based upon the 40-phon equal loudness contour to the signal. Thus all frequency components in the signal are weighted so that they make a contribution to the overall figure dependent upon their perceived loudness, rather than upon their actual intensity.

Although this scheme appears reasonable, when applied to speech, it takes no account of the importance of different frequencies to quality or intelligibility – only to their hearing perception. Additionally it does not account for absolute loudness, since the 40-phon curve really only describes the perception of a mythical average human to a 1 kHz signal at 40 dB$_{SPL}$.

More concerning is the fact that these historical and pioneering data were gathered based on single-tone signals rather than complex sounds, but are almost always applied to complex sounds. Still, in the absence of better methods, and scientific data indicating otherwise, A-weighting can be considered a reasonable approximation to many situations and applications outside the boundary of its known validity.

A-weighting has been ratified by the International Standards Organisation (ISO), as have ISO B- and C-weighting based on the shapes of the 70- and 100-phon curves respectively.

As an aside, dBA is commonly used for level measurements of 'nuisance' noise. However this only works if the audio frequencies being measured fit to the 40-phon curve. Infrasonic frequencies (2–32 Hz) and ultrasonic frequencies (> 20 kHz) both fall almost entirely outside the measurement domain. An example is given in [3] where householders near New York's JFK airport complained about excessive rumble from passing aeroplanes. This prompted tests for noise pollution which indicated that very little low-frequency noise was present. The rumble sound came from much higher frequency harmonics inducing a low-frequency perceived sound (we will see how this happens in Section 4.4.1). On an A-weighted scale the noise level measured would be low, whereas the sound experienced by the householders was loud.

---

The equal loudness contours as plotted in Figure 4.2 are the result of a number of factors in the human hearing system, one of which is the filtering (frequency selectivity) introduced by the pinna: *orthotelephonic gain*. The frequency distribution impinging on the eardrum differs when inner-ear headphones are used as opposed to loudspeakers, as the pinna provides around 5 dB gain at 2 kHz, 10 dB gain at 4 kHz and 2 dB gain at 8 kHz [8]. The filtering effect of the pinna below 500 Hz is negligible [3].

### 4.2.2    Cochlea echoes

When the ear is stimulated with tones, active processing produces components with differing frequencies and amplitudes from the stimulating tones [4]. For example, if only two tones ($f1$ and $f2$ with $f2 > f1$) are present then one may perceive a tone at a frequency of $2f1 - f2$. This effect is well-known to musicians who exploit it to induce musical harmony. Note that a similar effect is found when complex harmonic structure, with fundamental frequency removed is perceived as having a frequency equal to the fundamental (see Section 4.2.12).

### 4.2.3    Phase locking

As noted previously, neural excitation by the hair cells only occurs at the rarefaction part of the sound wave, at approximately fixed phase, although cells can vary considerably as to their individual phase positions. The average firing frequency for all cells will be of the correct frequency and correct phase. Some cells only fire every two, four or six cycles but this does not alter the overall firing rate [6]. Due to this cycle averaging process that the ear seems to use to distinguish between frequencies, it is possible for the ear to become attuned to a particular frequency, with hair cells firing in time with a rarefaction and not recovering until a short time later. Another rarefaction in this recovery period may be missed. The result being that a louder than usual amplitude will be required in order for a second tone (causing the additional rarefaction) to be heard [4].

This phase locking, as it is called, may well be part of the explanation for why a single tone will suppress tones of similar frequency and lower amplitude, part of the simultaneous masking phenomenon discussed in Section 4.2.8.

### 4.2.4    Signal processing

Cells in the auditory cortex are excited by a variety of acoustic signals, however cells do not in general respond to single tones, but require more complex sounds [6]. Natural processing in the human brain can detect tone start, tone stop, tone pulse, frequency slide up, frequency slide down, amplitude variation and noise burst conditions. One experiment has even determined the position of a group of brain cells that specialises in detecting 'kissing' sounds [4]. One wonders at the experimental methods employed.

### 4.2.5    Temporal integration

The ear's response with respect to time is highly nonlinear. For a tone duration below 200 ms, the intensity required for detection increases with decreasing duration, linearly proportional to the duration multiplied by the intensity required for detection of a constant tone of that frequency. Tones of longer duration – above about 500 ms – are detected irrespective of their duration, complexity and pattern [4].

In a similar way, periods of silence introduced into a constant tone are detectable to an extent which is dependent upon duration up to a duration exceeding about 200 ms.

### 4.2.6      Post-stimulatory auditory fatigue

After an abnormally loud sound the ear's response is reduced during a recovery period, after which it is returned to normal [3]. This is termed *temporary threshold shift* (TTS) [4]. The degree of TTS depends upon the intensity of the fatiguing stimulus, its duration, frequency and the recovery interval. It is also frequency specific, in that the TTS effect is distributed symmetrically about the frequency of the fatiguing tone and is limited to its immediate neighbourhood, but the actual frequency spread of the TTS curve is related to the absolute frequency. It is also related to the amplitude of the tone, and to the logarithm of the duration of the fatiguing tone (although the middle ear reflex muscle action reduces TTS for low frequencies, and a tone duration of over five minutes will produce no appreciable increase in TTS). When the fatiguing noise is broadband, TTS occurs mostly between 4 and 6 kHz, begins immediately and may still be noticeable up to 16 hours after the noise onset [9]. Tones louder than about 110 or $120\,\mathrm{dB}_{SPL}$ cause permanent hearing loss, but TTS is most prominent for amplitudes of 90 to $100\,\mathrm{dB}_{SPL}$.

### 4.2.7      Auditory adaptation

The response of a subject to steady-state tones will decline to a minimum over time, although an amplitude of about $30\,\mathrm{dB}_{SPL}$ is needed to trigger the effect. For the user this might be noted as a particular interfering sound becoming less noticeable with time. However it is worth noting that this effect appears to be highly subjective, with some subjects experiencing tones disappearing totally whilst others only experience a 3 dB or smaller reduction. The auditory system cannot adapt to truly broadband noise, and literature reports that high frequency tones are easier to adapt to than low frequency tones [4]. Despite these reports concerning broadband noise, it appears anecdotally that on long aeroplane journeys, an initial high level of cabin noise (which is subjectively quite broadband) becomes almost unnoticed by the end of a flight.

### 4.2.8      Masking

Masking deserves a chapter by itself, and indeed will form the basis of psychoacoustic modelling later in this book, but for now a basic introduction is sufficient. Masking in general is defined by the American standards agency as 'the process by which the threshold of audibility for one sound is raised by the presence of another sound' and 'the amount by which the threshold of audibility of sound is raised by the presence of another sound'. The frequency selectivity of the basilar membrane may be considered similar to the functionality provided by a bank of bandpass filters with a threshold of audibility in each filter being dependent upon the noise energy falling within its passband [10]. The filters each have similarly shaped responses with bandwidths approximately 100 Hz up to frequencies of about 1 kHz. Above this, bandwidth increases in a linear fashion with frequency up to a 3 kHz bandwidth at 10 kHz. Each 'filter' in the array is termed a critical band filter [11]. We will return to these filters later in Section 4.3.2.

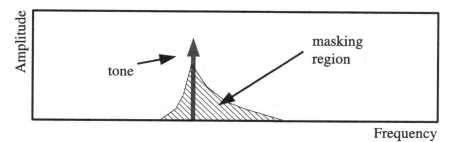

**Figure 4.3** Illustration of masking effect due to a single tone.

The effect of the critical band filters is that for a given tone having fixed frequency and amplitude, the sensitivity of the ear to tones of similar frequency is reduced. This is illustrated in Figure 4.3 which plots the frequency response of an artificial example tone, and overlays the masking area on top of this. While listening to the tone shown, any new tone which is introduced within the identified masking area, will be inaudible.

In general, once a louder tone has 'occupied' the sensors of one critical band filter, the same filter is less sensitive to other coincident sounds. Many researchers have attempted to derive logical models of this masking process, and these models exhibit a range of computational complexity and accuracy (some established models can be found in the literature, including the following: [9, 11, 12–16]).

As an example to illustrate the masking effect, let us create two pure tones in MATLAB using the tonegen() function from Section 2.7.1, ensuring that the lower frequency tone is only 20% of the amplitude of the louder one:

```
lo=0.2*tonegen(800, 8000, 2);
hi=tonegen(880, 8000, 2);
```

Next we will use sound() to replay the audio instead of soundsc() so that we can appreciate the differences in amplitude of the two tones. We will first listen to both tones alone, then we will listen to the two tones mixed together:

```
sound(lo/2, 8000);
sound(hi/2, 8000);
sound((lo+hi)/2, 8000);
```

Both of the individual tones can be heard when played alone, although the lower frequency tone is clearly quieter. However when replayed together the result is a slightly high tone exhibiting a slight warble. The low tone should be inaudible — masked by the louder tone.

One further interesting related point is that for sounds whose bandwidth falls entirely within one critical band, the intensity of that sound is independent of its bandwidth. However for sounds with bandwidth greater than one critical band, the intensity depends strongly on the proportion of the sound's bandwidth falling within one critical band.

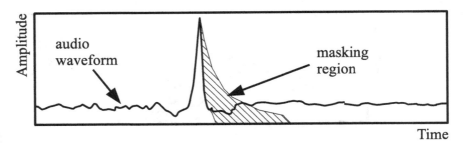

**Figure 4.4** Illustration of post-stimulatory masking in the time domain caused by a momentary loud sound.

Thus, in general, a complex sound having components in more critical bands sounds louder than a complex sound with components in fewer critical bands [4].

### 4.2.9    Co-modulation masking release

Although the amount of masking relies upon noise power, modulated noise (even when randomly modulated), whose bandwidth extends across more than one critical band, induces less masking as its bandwidth increases [5].

This phenomenon indicates that any correlation between critical bands can be utilised by the auditory system to enhance its performance (in this case, to reduce the effect of masking noise).

### 4.2.10    Non-simultaneous masking

Non-simultaneous or temporal masking can be subdivided into pre-stimulatory (backward) and post-stimulatory (forward) masking. In the former, masking effects occur *before* the onset of the masking tone, and in the latter, masking effects occur after the onset of the masking tone. Figure 4.4 provides an illustration of the effects of post-stimulatory masking. It plots the time domain waveform of some audio, and shows a masking region immediately following a loud sound, during which other sounds will not be audible.

Pre-stimulatory masking is notoriously difficult to measure, as it is highly dependent on the subject, and also on any previous exposure the subject has had to the test situation (which may cause the measurable effects to disappear completely) [4]. Normal practice in auditory testing is to 'warm up' with some dummy or calibration tests prior to the procedure beginning, which is not really possible when testing for pre-stimulatory masking.

Effects may occur for signals lasting up to a few hundred milliseconds, and persist for a similar length of time, with the rate of recovery (rate of reduction of masking effect) being higher for louder sounds. The duration of the masking tone determines the degree of masking up to about 20 ms, where the dependency levels out, and obviously the masking effect has frequency dependence based on the frequency of the masking tone [4,14].

## 4.2.11   Frequency discrimination

Frequency discrimination by humans is dependent upon absolute frequency, and to some extent amplitude, however an approximate figure would be 2 Hz for a 65 dB$_{SPL}$ signal at 1 kHz. Thus a tone of 1002 Hz can just be distinguished from a 1000 Hz tone by an average listener.

Frequency discrimination is related to pitch perception which decreases with increasing amplitude for tones below about 2 kHz, but increases with increasing amplitude for tones above about 4 kHz. Unless the subject is one of the 1% of the population capable of *absolute pitch* or *perfect pitch* then a pitch above 2.5 kHz, even when presented with a reference, cannot be discriminated. It is worth noting that due to this, tones of frequency greater than 5 kHz cannot evoke a sensation of melody [4].

We can easily test out frequency discrimination in MATLAB using the tonegen() function that was given in Section 2.7.1. We simply create and replay slightly different pure sinusoidal tones and see if there is any noticeable difference. In this case, let us use 1000 Hz and 1002 Hz tones of two seconds duration at 8 kHz:

```
t1=tonegen(1000, 8000, 2);
t2=tonegen(1002, 8000, 2);
soundsc(t1, 8000);
soundsc(t2, 8000);
```

For a more scientific test we would preferably present a sequence of such tones in an unknown sequence and ask listeners to identify which are higher and which are lower. Over many repetitions, a score significantly greater than guesswork (50% correct) would indicate that the tones can be discriminated.

## 4.2.12   Pitch of complex tones

A series of tones at 200, 400, 600, 800 Hz, . . . , evokes a pitch sensation of 200 Hz (as may be expected since this is the fundamental frequency in the set), however removal of the 200 Hz frequency does not affect the overall perceived tone. It does, however, affect the timbre or quality of the sound experienced. The perceived tone, when not actually present, is termed the residue [4].

This effect, which is known to musicians and composers, is very easy to demonstrate using MATLAB. First we create four pure sinusoidal tones of duration two seconds at an 8 kHz sample rate. We will use a fundamental of 196 Hz ($G_3$ on the musical scale) and its multiples:

```
t1=tonegen(196, 8000, 2);
t2=tonegen(196*2, 8000, 2);
t3=tonegen(196*3, 8000, 2);
```

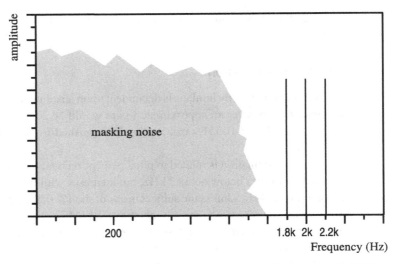

**Figure 4.5** An artificial arrangement causing 200 Hz to be heard, but where the 200 Hz tone, if present, would not be discernible due to heavy masking noise.

```
t4=tonegen(196*4, 8000, 2);
```

and then we will simply play back these, firstly a pure $G_3$. Remember the pitch of this tone, then we will play back the complex note both with and without the fundamental:

```
soundsc(t1, 8000);
soundsc(t1+t2+t3+t4, 8000);
soundsc(t2+t3+t4, 8000);
```

It should be noticeable that although the quality of the sound differs somewhat, the musical note being replayed is the same in each case.

Musically speaking, the fundamental frequency is the perceived note being played. Harmonics and other tones may be present, with timbre or quality changing greatly depending on the proportions. This is one of the defining characteristics of different musical instruments. Even when low frequency noise masks the perceived note (the residue frequency) such that it cannot be heard, the frequency of the residue is still perceived [17]. This is demonstrated in Figure 4.5, where the tones shown would evoke a response at 200 Hz, a frequency that if really present at that volume would be swamped by noise.

We find that both arithmetic and geometric relationships of single tones are discerned by the ear as being composite in some way.

**Table 4.1.** Binaural masking conditions and features (constructed from data presented in [18]).

| Noise condition | Signal condition | Processing gain |
|---|---|---|
| in phase | in phase | 0 dB (reference condition) |
| uncorrelated | out of phase | 3 dB |
| uncorrelated | in phase | 4 dB |
| out of phase | in phase | 6–13 dB |
| in phase | out of phase | 9–15 dB |

**Table 4.2.** Binaural masking conditions and features (constructed from data presented in [18]).

| | | Percentage of words correctly identified | | |
|---|---|---|---|---|
| | | Phasic | Random | Antiphasic |
| stereo | phasic | 18 | 27 | 35 |
| | antiphasic | 43 | 27 | 16 |
| mono | right ear | 30 | 13 | 20 |
| | left ear | 18 | 8 | 15 |

## 4.2.13 Binaural masking

The human brain can correlate the signals received by two ears to provide a processing gain. This is measured by presenting a signal of noise plus tone to both ears, and adjusting the tone level until it is just masked by the noise. From this situation a change is made, such as inverting the tone phase to one ear, and the tone amplitude is again adjusted until just barely audible. The difference in amplitude of the tone in the second case reveals the processing gain. Table 4.1 (constructed from data presented in [4,9]) summarises the results under different conditions.

When the signal is speech, intelligibility is reduced in anti-phasic (uncorrelated noise, out of phase speech) conditions over phasic conditions (uncorrelated noise, speech in phase) as revealed in Table 4.2.

The effects of binaural masking are greatest at lower frequencies, but depend upon the frequency distribution of both the test signal and the noise.

Note that *binaural* is a term used by auditory professionals and researchers. It means 'pertaining to both ears', and is frequently confused with *stereo*. Stereo actually refers to sound recorded for multiple loudspeakers, which is designed to be replayed several metres in front of a listener, and generally enjoyed by both ears. Users of headphones are thus exposed to a significantly different sound field to those listening with loudspeakers – a fact which is not normally exploited by audio equipment manufacturers.

## 4.2.14      Mistuning of harmonics

Two complex musical tones are perceived as separate when they have different fundamental frequencies, however the hearing process is capable of dealing with slight mistuning of certain components, so almost-equal fundamentals can sometimes be perceived as being equal.

Again, MATLAB can be used to demonstrate this effect. The following section of code creates a complex musical chord, in this case an $A_4$, plus the notes a third and an eighth of an octave above this (refer to Infobox 2.5: Musical notes on page 33 for an explanation of the one-twelfth power used below):

```
note=440;
t1=tonegen(note, 8000, 1);
t2=tonegen(note*2^(3/12), 8000, 1);
t3=tonegen(note*2^(8/12), 8000, 1);
```

When replayed, this makes an obvious musical chord:

```
soundsc(t1+t2+t3, 8000);
```

Next, we will mistune the centre note in the chord by a factor of 5% and then listen to the result:

```
m2=tonegen(note*1.05*2^(3/12), 8000, 1);
soundsc(t1+m2+t3, 8000);
```

The resulting sound should still be perceived as a fairly pleasant-sounding musical chord, with a different quality to the correctly tuned chord, but musically still compatible, to the relief of piano tuners everywhere.

In general, a slight mistuning of one harmonic will result in a perception of reduced amplitude, until the degree of mistuning becomes such that the harmonic is perceived as a tone in its own right. Again, the effects depend on duration, amplitude and absolute frequency (as well as person-to-person differences), but a rule of thumb is that 400 ms long tones must be mistuned by over 3% for them to be heard separately [19]. Note that this is not the same effect at all as the beats frequency caused by two interfering tones.

## 4.2.15      The precedence effect

The *Haas* or precedence effect ensures that if similar versions of a sound reach an observer at slightly delayed times, then the observer will hear the first signal but suppress the subsequent versions [3]. This effect, only acting on signals reaching the ear within 50 ms of each other, explains why we can still understand speech in an environment containing multiple echoes (such as a small room). The first sound to reach the observer will be heard in preference to further sounds even if the secondary sounds are up to 10 dB louder.

Once echoes reach an observer with time delays of more than about 65 ms (corresponding to a distance of approximately 20 m in air), they will be perceived as being distinct echoes. For this reason, echoes are only noticeable in large rooms – they are present in small rooms but we are unable to hear them.

MATLAB can be used to construct a demo of the Haas effect. First record a word or two of clear speech into MATLAB, then use the following code to listen to the speech replayed with progressively longer unattenuated echoes. Fs is set to the sample rate used to record the speech, and the echo starts at 10 ms, then increases in steps of 20 ms up to 100 ms:

```
audio=reshape(audio,1,length(audio));
for echo=0.01:0.020:0.1
  pad=zeros(1,fix(Fs*echo));
  input('Press any key to hear next echo');
  soundsc([audio,pad]+[pad,audio],Fs);
end
```

Note that the reshape() function is used to ensure that the audio vector is $(1 \times N)$ rather than $(N \times 1)$ in dimension, so it can be combined with the variable sized padding.

One useful rule of thumb when working with speech is that echoes of 100 ms or longer in two-way conversational speech, such as during a telephone call, will become annoying and distracting to callers. This is a major quality factor for telephony systems, and presents a hard-to-meet target for IP-based Internet telephony systems such as Skype which use digital processing and relatively slow or congested communications networks.

## 4.2.16   Speech perception

Evidence exists to indicate that the human aural system processes speech in a completely different way from other non-similar sounds [4]. One experiment which uses sinewave speech to demonstrate this, was discussed in the introduction to Chapter 3. Whatever the exact cause, certain brain mechanisms cause a listener to perceive two sentences of speech as being similar when the physical differences between them may be very large (for example, same speaker, same words, different acoustic situation). Two examples of one speaker saying the same words in two different environments would be judged by listeners as similar, irrespective of possibly major physical differences in amplitude, timing, pitch, and so on. Speech can be detected and understood when the noise-to-signal power is such that single tones would be inaudible.

Despite the evidence suggesting that speech processing in the brain is separate from sound processing, speech perception still suffers from simultaneous and non-simultaneous masking, binaural masking and auditory adaptation (indeed this may be more pronounced: monotonic speech rapidly leads to drowsiness or lack of interest). Speech perception interlinks speech production with the psychological aspects of the communication channel, the subject matter and the state of mind of the recipient [3].

## 4.3     Amplitude and frequency models

Attempts to classify and regularise the human hearing system have been made for many years. Most of these have revolved around the analysis of frequency and amplitude effects, and in particular the measurements of loudness, and frequency selectivity discussed below.

### 4.3.1     Loudness

Backtracking from the equal loudness contours of Section 4.2.1, it is fairly evident that humans do not measure the concept of loudness in the same way as a physical measurement. From our discussions previously we have seen that people interpret amplitude approximately logarithmically.

In fact, several researchers indicate that perceived loudness is proportional to physical amplitude raised to the power of 0.3 [4]. It is likely that there are upper and lower bounds where the relationship no longer holds true, nevertheless across the main portion of the hearing range the relationship is defined by a constant multiplier that varies from person to person. The measure of loudness is the *sone* although this is rarely used in current literature. One sone is the loudness of a 1 kHz tone at 40 dB$_{SPL}$, with every 10 dB increase in amplitude increasing the loudness by one sone (i.e. a 50 dB$_{SPL}$ 1 kHz tone has a loudness of two sones).

### 4.3.2     The Bark scale

Since the hearing process is often considered to derive from some set of bandpass filters, and the resultant processing within these, researchers attempted to identify and characterise these filters [4,10,11]. Known as *critical band filters*, each has similar shape but different bandwidth, centre frequency, and amplitude weighting. The different amplitude weightings contribute to the sone loudness scale (Section 4.3.1), as well as the equal loudness contours of Section 4.2.1 – both of which originally described the loudness of single tones only.

More generally, the loudness and frequency selectivity of sounds depend on the bands within which they fall. In particular this mechanism can be used to explain the masking effects of Section 4.2.8: a loud tone in one critical band will occupy that band and prevent a quieter tone in the same band from being heard. Moving the quieter tone gradually away from the louder tone until it enters a neighbouring critical band will result in it becoming audible once more. The *Bark*[1] scale is one way to express this relationship [20].

The first few lower critical bands are shown in Table 4.3 where they are numbered according to the Bark scale. The scale is arranged so that a unit change in Bark

---

[1] Since the scale is named after a person, Heinrich Barkhausen, for his early work on loudness perception, the name Bark should be capitalised to distinguish it from the sound a dog makes.

**Table 4.3.** Critical bands and corresponding centre frequencies.

| Critical band (Bark) | Lower cutoff frequency (Hz) |
|:---:|:---:|
| 1 | 100 |
| 2 | 204 |
| 3 | 313 |
| 4 | 430 |
| 5 | 560 |
| 6 | 705 |
| 7 | 870 |
| 8 | 1059 |
| 9 | 1278 |
| 10 | 1532 |
| 11 | 1828 |
| 12 | 2176 |
| 13 | 2584 |
| 14 | 3065 |
| 15 | 3630 |

corresponds to a perceived unit change in frequency effect by listeners. It is therefore a psychoacoustically-relevant frequency scale, but as always we need to remember that this table describes the mythical 'standard human'. In real life different listeners will have slightly different hearing characteristics.

Denoting a Bark unit as $\Omega$, and the angular frequency as $\omega$, then Hermansky [21] defines the Bark in the following way:

$$\Omega(\omega) = 6 \log \left( \omega/1200\pi + \sqrt{(\omega/1200\pi)^2 + 1} \right). \tag{4.1}$$

Simple MATLAB functions to convert bidirectionally between frequencies in Hz and Bark are given below:

```
function [bark]=f2bark(hz)
  cn=2*pi*hz/(1200*pi);
  bark=6*log(cn+(cn^2+1)^0.5);

function [hz]=bark2f(bark)
  hz=600*sinh(bark/6);
```

One final note of caution is to beware of alternative definitions of the Bark. There are at least three separate definitions of the simple mapping between Hz and Bark in use by research authors worldwide. Exactly which one is in use is not particularly important, since all are relative, and all map to real frequencies with a similar shaped representation, but it is critically important to be consistent and not combine or confuse the different definitions.

## 4.4          **Psychoacoustic processing**

The use of psychoacoustic criteria to improve communications systems, or rather to target the available resources to subjectively more important areas, is now common. Many telephone communications systems use A-law compression. Philips and Sony have respectively produced the DCC (digital compact cassette) and the MiniDisc formats which both make extensive use of equal loudness contours, and masking information to compress high quality audio [22]. Whilst neither of these were runaway market successes, they introduced psychoacoustics to the music industry, and paved the way for solid state music players such as the Creative Technologies Zen micro, Apple iPod and various devices from iRiver, Philips, and others too numerous to mention.

Most of these devices use the popular MP3 compression format (although its successor, MP4, and several proprietary alternatives exist). It is important to note that all of these music compression methods have something in common: they all use psychoacoustic criteria. All take account of masking thresholds, some consider harmonic relationships, and others exploit binaural masking.

In the remainder of this section, several promising psychoacoustic techniques with potential for use in speech compression are discussed. Not all of these have yet been exploited commercially, and few are in mainstream research.

### 4.4.1          Tone induction

Knowledge of the residue effect (Section 4.2.12) allows one to induce a low frequency tone by the addition of higher frequency tones. Only three or so high frequency tones are required, and the technique is useful where it would otherwise be necessary to directly add an extremely loud low frequency tone in order to be heard above a low frequency noise. One application of this is in the construction of an artificial speech formant (formants apparently being regarded by Klatt, a leading audio and speech researcher, as being the most important aspect of speech recognition [4]). This application has not yet been substantiated. Critical to its success is the location within the auditory cortex of the mechanisms causing the residue effect. If these mechanisms occur posterior to the mechanisms of speech recognition, then formant induction is unlikely to be successful, and the listener would simply hear speech degraded by high frequency tones. Another application is in the induction of bass notes using small speaker cones. Generally, for low frequency bass notes to be reproduced, a large loudspeaker cone is needed. However such items are expensive, unwieldy, and power hungry. This method could potentially be used to allow smaller speakers to induce the kind of sound normally requiring larger, more costly, speakers.

### 4.4.2          Sound strengthening

As stated in Section 4.2.12, adding harmonics to a complex tone, or adding geometrically related tones, does not change the perceived frequency of a sound, but does change the amplitude (firstly by allocating more signal power to the sound, and secondly by spreading the components into more critical bands). This relationship allows the 'strengthening' of a sound by the addition of related frequency components without causing undue concentration of energy at specific frequencies, as would be the case with direct amplification of those fundamental frequencies to achieve a perceptually similar result.

This technique may allow a speech formant, obscured by localised noise, to be perceptually strengthened [23].

### 4.4.3 Temporal masking release

Post-stimulatory temporal masking (see Section 4.2.10), causing sounds occurring just after a loud tone or pulse to be perceived less loud, may be negated by detecting such events and applying an initial increase, followed by a decaying level of amplification after it. A knowledge of the exact amplitude at the ear of a listener is probably a necessary requirement for this processing. In a similar way for much louder sounds, allowances may be made for TTS (temporary threshold shift – see Section 4.2.6).

### 4.4.4 Masking and two-tone suppression

If a wanted tone lies close to a masking tone of higher amplitude, it is known that the wanted tone will be masked (see Section 4.2.8). To alleviate this situation, the wanted tone could be amplified, strengthened by the addition of harmonics, or simply shifted slightly. If the wanted tone is harmonically related to other tones, we know that the shift may not be more than about 3% [4]. Techniques already exist for predicting the effects of masking for the purpose of enhancement, many of which are reported to perform fairly well [14,15,20,24,25].

### 4.4.5 Use of correlated noise

Remember that modulated noise, present in more than one critical band, allows the auditory system to use a correlation process to reduce its effects (refer to Section 4.2.9). In this way, we can conjecture that a frequency shifted, reduced amplitude version of that noise induced elsewhere in an unimportant part of the frequency spectrum should reduce its effect.

### 4.4.6 Binaural masking

As noted in Section 4.2.13, judicious control of the signal fed to each ear of a listener may help to improve the perceived signal-to-noise level. For speech systems contaminated with additive white Gaussian noise, it appears that applying out-of-phase speech to each ear, whilst subject to the same noise field, can reduce the effect of the noise.

## 4.5     Auditory scene analysis

Auditory scene analysis (ASA) [26] describes the psychoacoustic and psychological processing by the human auditory system that it uses to deal with and interpret complex mixtures of sound. This topic is clearly the result of some type of analytical processes occurring within the brain.

As an illustration, consider an environment when a person is talking, a dog is barking and a police siren are heard concurrently. Each of these sounds arrives at the ears of a listener simultaneously, and thoroughly mixed together. A computer, analysing this sound mixture, would have great trouble in determining that there were three separate sound sources, and even if it could determine this, may not readily be able to 'tune in' to each of those sounds in the way a human can.

This tuning in process comes into play when participating in a conversation in a crowded room filled with multi-speaker babble. Tens or even hundreds of separate conversations may be simultaneously in progress, and yet most people will be able to tune out much of the surrounding babble and conduct a conversation with the person next to them. This is true even when the particular speech they are listening to is of lower amplitude than the interfering sounds.

Imagine yourself in such a situation. You are straining to hear the person opposite you who is barely at the edge of audibility, and yet somehow a conversation is possible in all that noise. It is true that occasionally a particular remark or name, or perhaps an overloud laugh from a nearby speaker, may capture your attention, thus intruding on your conversation and prompting a request to repeat something. However you are able in some way to tune in to particular conversations occurring around you while tuning out much of the other noise.

Sometimes visual clues help to maintain that focus – if you can see the mouth of the person speaking, the movement and shape of the mouth are interpreted by the human visual system, providing clues to the auditory system. This was illustrated in a famous experiment in which listeners were played /b/ and /m/ sounds whilst simultaneously watching videos of people saying /b/ or /m/.

When the correct video was played simultaneously with a replayed sound, listeners correctly identified the spoken phoneme. However when the video of /b/ was replayed alongside the audio for /m/, most listeners reported that they had heard /b/. The image of the mouth closure at the start of the /b/ sound was sufficient to unconsciously convince listeners that they had heard /b/ rather than /m/. This is known as the McGurk effect, and is an example of the human visual system assisting (and indeed overriding) the human auditory system (HAS).

Interested readers may wish to refer to an article in *New Scientist* of 22 September 2007 'Mind Tricks: Six Ways to Explore your Brain' in which a McGurk effect experiment, and several similar experiments are described.

All of these are illustrations of the way in which the HAS deals with, or interprets sounds. In order to begin characterising these effects, researchers have identified several

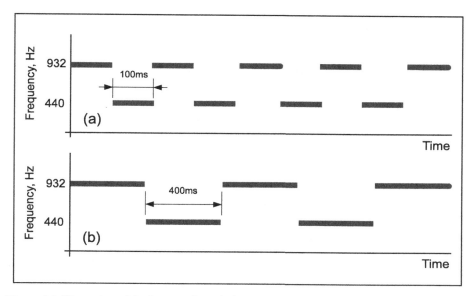

**Figure 4.6** Illustration of the feature of proximity. Two sequences of alternating tones are shown. In plot (a) the absolute time difference between like tones is short, thus they exhibit closer proximity compared to the larger time difference between like tones in plot (b).

basic principles [26,27] which relate complex sounds and sequences to perception, and which we will consider in the following subsections.

## 4.5.1   Proximity

Sounds of similar proximity are those that are close in terms of amplitude, pitch, duration, timbre, and so on. There is definitely a tendency for the brain to classify similar sounds as belonging to the same parts of an auditory scene.

For example, when listening to a recording of a wind quartet, the brain would probably be correct in assuming that each oboe-like sound originates from the same oboe. If, in fact, two oboes are playing, then that may only become evident when both play different notes simultaneously.

We shall use MATLAB code to generate two sound sequences, each consisting of two alternating tones. One first sequence plays the tones slowly, the other sequence plays them quickly. The sequences are illustrated in Figure 4.6 showing a closer proximity in the upper plot compared to the lower, due to increased gap between similar frequencies in the latter. A rather simple MATLAB demonstration, using the tonegen function of Section 2.7.1, can illustrate this point:

```
ss=0.1;      %short sound length in seconds
ls=0.4;      %long sound length in seconds
Fs=8000;     %sample rate
short_a=tonegen(440,Fs,ss); %concert A4
```

```
short_b=tonegen(932,Fs,ss); %B5
long_a=tonegen(440,Fs,ls);
long_b=tonegen(932,Fs,ls);
%Construct alternating sounds in a single matrix
short_mat=[short_a, short_b];
long_mat=[long_a, long_b];
%Repeat this matrix several times
long=repmat(long_mat,1,3);
short=repmat(short_mat,1,12);
```

The two tone sequences reside in vectors long and short respectively. Let us begin by listening to the long duration tones:

```
soundsc(long, Fs);
```

Most likely this is heard as a single sequence consisting of two tones. However, when listening to the faster sequence of short duration tones, it is more likely that listeners notice two streams of sound – one high frequency, and one low frequency, but both are pulsed:

```
soundsc(short, Fs);
```

To explain, both sequences exhibit no gaps between tones, thus there is close temporal proximity between an $A_4$ and the following $B_5$. Although there is not a great pitch proximity between the $A_4$ and the $B_5$ (which is slightly over an octave higher), in the slower sequence, the close temporal proximity dominates, leading the brain to associate the tones together into one slow string.

In the faster sequence, however, the proximity of the like notes is increased – they are now only 100 ms apart – and thus the proximity of one $A_4$ to the next is increased over the slow sequence. Likewise, each $B_5$ is nearer to the next one. In this recording, the close tonal proximity dominates, and thus two separate streams are heard. One is of beeping $A_4$ notes and one is of beeping $B_5$ notes.

## 4.5.2     Closure

The closure principle allows the brain to fill any auditory 'gaps' to match its expectations regarding continuity, origin, and so on. This is also known as auditory induction since our brains do not just *deduce* the presence of missing audio, but actually *induce* a replacement to cover the missing gaps. It seems our brains dislike sounds that do not fit into an ongoing scene interpretation, to the extent that additional material will be created to bring the user perceived audio in line with brain expectations. The same process is involved with images, and is the basis for many optical illusions, from impossible three-dimensional shapes, to discovering hidden faces within abstract images.

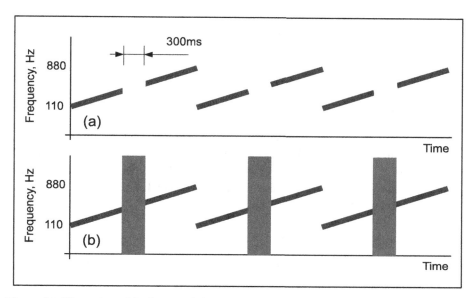

**Figure 4.7** Illustration of the feature of closure using steadily rising tonal frequencies. In the first sequence (a), these are interrupted by gaps, whereas in the second sequence (b), the rising tones are interrupted by white noise.

In audio terms, this process of filling in expected information happens commonly during speech communications, especially when carried out in the presence of interfering noise. At a higher language level, this would be one of the reasons why context is so important to speech intelligibility (discussed in Section 3.3.4).

In MATLAB, we can demonstrate this effect by again using the freqgen function of Section 2.7.3 to create a sound comprising of rising tones, and then to insert either gaps or noise portions into the sound. The pattern created is shown in Figure 4.7, using the following MATLAB code:

```
gs=0.30;        %gap/noise length in seconds
ls=1.50;        %length of sound in seconds
Fs=8000;        %sample rate
fr=110;
to=880;
gap_len=Fs*gs;
au_len=Fs*ls;
gap=zeros(1,gap_len);
noise=rand(1,gap_len);
%Make a steadily rising note
note_f=[fr:(to-fr)/(au_len-1):to];
au=freqgen(note_f,Fs);
au_gap=au;
au_noise=au;
```

```
%Now put the gaps 1/2 way up
au_gap(au_len/2+1:au_len/2+gap_len)=gap;
au_noise(au_len/2+1:au_len/2+gap_len)=noise;
%Now repeat several times
au_gap=repmat(au_gap,1,3);
au_noise=repmat(au_noise,1,3);
```

Replaying the noisy version first, listeners will hear a steady rising sinewave, moving from 110 Hz to 880 Hz. During the rising period, there will be three periods of loud white noise. The interpretation by most listeners is that the sinewave is continuous, but parts of it cannot be heard because of the noise. Try and decide whether you think the sinewave is continuous:

```
soundsc(au_noise, Fs);
```

By contrast, when replacing the noise periods by a gap of silence, the discontinuity in the sinewave is immediately obvious:

```
soundsc(au_gap, Fs);
```

By and large, humans are fairly sensitive to discontinuities – whether audio, visual (edges) or even tactile and thus the fact that two discontinuities can be hidden by white noise of the same average amplitude is significant. As a final check, we can compare how audible the sinewave would really be if it were replayed simultaneously with the noise instead of being replaced by the noise:

```
soundsc(repmat(au,1,3)+au_noise, Fs)
```

The answer is that the sinewave is clearly audible even in the presence of that noise. The HAS interpreted the first playback as being a continuous sinewave, even when the sinewave was not audible, but would have been were it actually present. This is clearly an illustration of the brain filling in gaps in its interpretation of the events which it witnesses.

Interestingly, GSM and several other speech coders introduce a sound known as *comfort noise*, a pseudo-random white noise sound, at times when silence would otherwise be present. This is done to improve the perception of listeners, based on observations which showed that listeners were discomforted by periods of total silence during a voice communications session. Perhaps, in line with our experiment to test the effects of closure using periods of noise, another advantage of comfort noise is the ability to 'blind' the HAS to otherwise disturbing and obvious waveform discontinuities in a speech coder output.

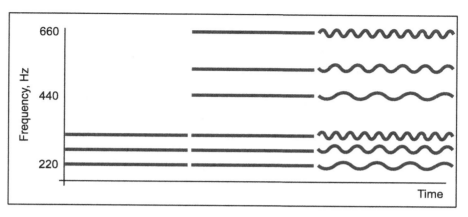

**Figure 4.8** Illustration of the feature of common fate by comparing three sections of audio tones. The first section reproduces a pleasing note generated from three related sinewaves of different frequencies. The second section extends each of the three fundamentals of the original sound with their first harmonics. The third section modulates each of the three tones plus their first harmonic. The three modulating frequencies are unrelated.

### 4.5.3 Common fate

When groups of tones or noises start, stop or fluctuate together, they are more likely to be interpreted as being either part of a combined sound, or at least having a common source. As an example, a human voice actually consists of many component tones which are modulated by lung power, vocal tract shape, lip closure and so on, to create speech. It is likely that the ability to cluster sounds together by 'common fate' is the mechanism by which we interpret many sounds heard simultaneously as being speech from a single mouth. In noisy environments, or in the case of multi-speaker babble, common fate may well be the dominant mechanism by which we can separate out the sounds of particular individuals who are speaking simultaneously.

The same effect, and outcome, can be heard in an orchestra, where several musical instruments, even playing the same fundamental note simultaneously, can be interpreted by the HAS as being played separately (and the more different the instrument is, the more unique is the modulation, and thus the easier they are to distinguish, for example a violin and a bassoon are easier to distinguish than two bassoons playing together).

Let us attempt to demonstrate this effect in MATLAB by creating three sets of combined audio notes for playback, as illustrated in Figure 4.8. First of all we will define three sinewave frequencies of $a$, $b$ and $c$, related by a power of $2^{(1/12)}$ so that they sound pleasant (see Infobox 2.5 on page 33 which describes the frequency relationship of musical notes). We will then use the tonegen function of Section 2.7.1 to create three sinewaves with these frequencies, and also three sinewaves of double the frequency:

```
dur=1.2;
Fs=8000;
a=220;
```

```
b=a*2^(3/12);
c=a*2^(7/12);
sa=tonegen(a,  Fs,  dur);
sb=tonegen(b,  Fs,  dur);
sc=tonegen(c,  Fs,  dur);
sa2=tonegen(a*2,  Fs,  dur);
sb2=tonegen(b*2,  Fs,  dur);
sc2=tonegen(c*2,  Fs,  dur);
```

Next, two sounds are defined which in turn mix together these notes:

```
sound1=sa+sb+sc;
sound2=sound1+sa2+sb2+sc2;
```

Now, three different modulation patterns are created, again using the tonegen function, but at much lower frequencies of 7, 27 and 51 Hz, chosen arbitrarily to not be harmonically related to either the original tones, or to each other. These are then used to modulate the various original tones:

```
mod1=tonegen(7,  Fs,  dur);
mod2=tonegen(27,  Fs,  dur);
mod3=tonegen(51,  Fs,  dur);
am=mod1.*(sa+sa2);
bm=mod2.*(sb+sb2);
cm=mod3.*(sc+sc2);
```

Finally, a short gap is defined for the third sound to accentuate the common starting point of three sound components, which are combined into sound3, and finally placed into a composite vector for replaying:

```
gap=zeros(1,Fs*0.05);
sound3=[am,gap,gap]+[gap,bm,gap]+[gap,gap,cm];
soundsc([sound1,sound2,sound3],  Fs)
```

A listener exposed to the sounds created should hear, in turn, three segments of audio lasting approximately 1.2 s each. Firstly a pleasing musical chord consisting of three harmonics is heard. Then this musical chord is augmented with some harmonically related higher frequencies. For both of these segments, the perception is of a single musical chord being produced. However the final segment is rather discordant, and appears to consist of three separate audible components.

Because each of the three harmonic notes from the second segment are now modulated differently, the brain no longer considers them as being from the same source, but rather from different sources. It thus separates them in its perceptual space.

To really show that this interpretation is not simply a side-effect of the small gap that was inserted, the reader is encouraged to repeat the experiment, but modulate bm and cm in the same way as am, namely:

```
am=mod1.*(sa+sa2);
bm=mod1.*(sb+sb2);
cm=mod1.*(sc+sc2);
```

With all subsequent steps repeated as previously, the listener will hear, instead of a discordant mixture of three notes, a warbling single chord. This is to be expected since now each of the three harmonic notes shares a common fate (a common modulation).

## 4.5.4 Good continuation

Generally, sounds in nature do not start, stop or switch frequency instantaneously. There will normally be some amplitude attack at the beginning (ramping up from zero) and decay (ramping down to zero) at the end of a sound. Similarly, sounds in nature normally do not just switch frequency instantaneously, they slide either quickly or slowly.

Both of these properties are due to the physical methods of creation of such sounds. Changes in frequency are caused by changes in the shape or geometry of the producing object, such changes would be driven by muscles in humans and animals, which move relatively slowly and gradually in audio terms, so that at least some intermediate frequencies will be present between two extremes. Similarly, sounds are conveyed by the movement of air, and air movements can be sharp but not completely instantaneous. They must at first build up, and later die away.

So when presented by complex sets of sound, the human brain tends to classify sounds that are connected in some way by these gradual changes as coming from a single source. By contrast, sounds which do not blend together are more likely to be regarded by the brain as coming from separate sources. Researchers have termed this effect the good continuation of sounds, although it could perhaps better be described as being the *connectedness* of individual sounds.

Musical instruments are similar. In particular their notes always exhibit some form of amplitude attack and decay, however notes do not always glide into one another.

Computers provide us the ability to generate almost any arbitrary set of sounds, with whatever features we may require. Computer generated sounds can start, stop and switch frequency instantaneously if we wish, or could be tailored to show some connectedness.

In the following MATLAB code, we will aim to demonstrate the effect of the good continuation property by constructing two sequences of sound. The first, shown in Figure 4.9(a), has two alternating notes joined by a sliding frequency, or *glissando*, while the second, in plot (b), presents the same two notes without any glissando.

Listeners of both sequences should consider that the first could conceivably be produced by a single source which slides upwards and downwards in frequency. The second, by contrast, is more likely to be interpreted as two separate, single frequency sound sources that alternate in producing a sound.

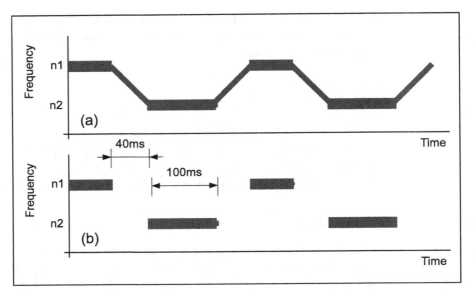

**Figure 4.9** Illustration of the feature of good continuation between notes in plot (a) which is missing in plot (b).

In MATLAB, first we will produce the two note sequences, sa and sb, using the freqgen() function, with the latter sequence exhibiting the glissando:

```
Fs=8000;
n1=832;
n2=350;
d1=0.1*Fs;
dm=0.04*Fs;
d2=0.1*Fs;
a=[n1*ones(1,d1), zeros(1,dm), n2*ones(1,d2), zeros(1,dm)];
b=[n1*ones(1,d1), n1-[1:dm]*(n1-n2)/(dm), n2*ones(1,d2),
    n2+[1:dm]*(n1-n2)/(dm)];
sa=freqgen(a,Fs);
sb=freqgen(b,Fs);
```

We could plot the two frequency tracks, a and b, and see that they are very similar to the plots of Figure 4.9. Next we will perform some amplitude shaping by amplifying the notes at the expense of the glissando. This helps to present the glissando as a secondary feature rather than as the focus of the sounds:

```
amp=0.4+sign(a)/2;
sa=sa.*amp;
sb=sb.*amp;
```

Finally, we can replay several repetitions of the discontinuous sound:

```
soundsc(repmat(sa,1,8))
```

and follow that with the connected sounds:

```
soundsc(repmat(sb,1,8))
```

## 4.6     Summary

In this chapter we studied human hearing, beginning with the physical structure of the ear, and the processes occurring within it. We then discussed several counter-intuitive aspects of hearing, where the human auditory system fails to behave in the predictable way that an examination of the physics of sound would suggest.

This departure from a purely physical interpretation of sound, named psychoacoustics, was characterised through several examples, including MATLAB programs to demonstrate the effect in many cases. Finally, the entire topic of auditory scene analysis was discussed in an attempt to formulate a conceptual model describing the operation of the human auditory system in interpreting complex mixtures of sounds presented to it.

We will now depart from a discussion of the human auditory system to discuss speech communications in Chapter 5, returning to methods of automatic sound analysis by computer in Chapter 6, and finally back to several important psychoacoustics topics in Chapter 7.

## Bibliography

- *An Introduction to the Psychology of Hearing*
  B. C. J. Moore (Academic Press, 4th edition 1997)
  This book is highly recommended as an introduction to hearing research, in any of its four editions. Not only is it a well-written scholarly work, but it covers almost all aspects of hearing which may relate to computer handling of speech and music. It does not cover any processing topics, but is purely related to human experience, psychology and physiology.

- *The Handbook of Hearing and the Effects of Noise*
  K. D. Kryter (Academic Press, 1994)

- *Hearing (Handbook of Perception and Cognition)*
  B. C. J. Moore (Academic Press, 2nd edition 1995)

- *Speech Intelligibility and Speaker Recognition*
  Ed. M. E. Hawley (Halsted Press/Dowden Hutchinson and Ross, 1977)

- *The Psychophysics of Speech Perception*
  Ed. M. E. H. Schouten (NATO Science Series D, Martinus Nijhoff publishers/ Springer, 1987)

- *Auditory Scene Analysis: Grouping Principles (citing source [27])*
  www.ni.cs.tu-berlin.de/hendrik/research/asa.html

- *Auditory Scene Analysis*
  A. S. Bregman (MIT Press, 1990)

# References

[1]   R. W. Schaefer and L. R. Rabiner. Digital representation of speech signals. *Proc. IEEE*, **63**(4): 662–677, 1975.

[2]   W. G. Sears. *Anatomy and Physiology for Nurses and Students of Human Biology*. Arnold, 4th edition, 1967.

[3]   F. White. *Our Acoustic Environment*. John Wiley, 1976.

[4]   B. C. J. Moore. *An Introduction to the Psychology of Hearing*. Academic Press, 1992.

[5]   R. Duncan-Luce. *Sound and Hearing: A Conceptual Introduction*. Lawrence Erlbaum Associates, 1993.

[6]   W. F. Ganong. *Review of Medical Physiology*. Lange Medical Publications, 9th edition, 1979.

[7]   L. L. Beranek. The design of speech communications systems. *Proc. IRE*, pages 880–890, September 1947.

[8]   K. Kryter. *The Effects of Noise on Man*. Academic Press, 2nd edition, 1985.

[9]   W. Tempest, editor. *The Noise Handbook*. Academic Press, 1985.

[10]  P. J. Blamey, R. C. Dowell, and G. M. Clark. Acoustic parameters measured by a formant estimating speech processor for a multiple-channel cochlear implant. *J. Acoustical Soc. America*, **82**(1): 38–47, 1987.

[11]  M. R. Schroeder, B. S. Atal, and J. L. Hall. Optimizing digital speech coders by exploiting masking properties of the human ear. *J. Acoustical Soc. America*, **66**(6): 1647–, 1979.

[12]  Y. M. Cheng and D. O'Shaughnessy. Speech enhancement based conceptually on auditory evidence. *Proc. Int. Conf. on Acoustics, Speech and Signal Processing*, Vol. 2 pages 961–963, 1991.

[13]  Y. M. Cheng and D. O'Shaughnessy. Speech enhancement based conceptually on auditory evidence. *IEEE Trans. Signal Proc.*, **39**(9): 1943–1954, 1991.

[14]  Y. Gao, T. Huang, and J. P. Haton. Central auditory model for spectral processing. *Proc. Int. Conf. on Acoustics, Speech and Signal Processing*, Vol. 2 pages 704–707, 1993.

[15]  M. Mourjopoulos, J. Tsoukalas, and D. Paraskevas. Speech enhancement using psychoacoustic criteria. In *Proc. Int. Conf. on Acoustics, Speech and Signal Processing*, pages 359–362, 1991.

[16]  N. Virag. Speech enhancement based on masking properties of the auditory system. *Proc. Int. Conf. on Acoustics, Speech and Signal Processing*, Vol. 1 pages 796–799, 1995.

[17]  J. C. R. Licklider. Auditory feature analysis. In *Information Theory*. Academic Press, 1956.

[18]  K. Kryter. *The Handbook of Hearing and the Effects of Noise*. Academic Press, 1994.

[19]  C. R. Darwin and R. B. Gardner. Mistuning a harmonic of a vowel: Grouping and phase effects on vowel quality. *J. Acoustical Soc. America*, **79**: 838–845, 1986.

[20]  D. Sen, D. H. Irving, and W. H. Holmes. Use of an auditory model to improve speech coders. *Proc. Int. Conf. on Acoustics, Speech and Signal Processing*, Vol. 2 pages 411–415, 1993.

[21]  H. Hermansky. Perceptual linear predictive (PLP) analysis of speech. *J. Acoustical Soc. America*, **87**(4): 1738–1752, 1990.

[22]  "ISO/MPEG – Audio Standard layers". Editorial pages. *Sound Studio Magazine*, pages 40–41, July 1992.

[23]  J. I. Alcantera, G. J. Dooley, P. J. Blamey, and P. M. Seligman. Preliminary evaluation of a formant enhancement algorithm on the perception of speech in noise for normally hearing listeners. *J. Audiology*, **33**(1): 15–24, 1994.

[24]  A. Azirani, R. Jeannes, and G. Faucon. Optimizing speech enhancement by exploiting masking properties of the human ear. *Proc. Int. Conf. on Acoustics, Speech and Signal Processing*, Vol. 1 pages 800–803, 1995.

[25]  R. E. P. Dowling and L. F. Turner. Modelling the detectability of changes in auditory signals. *Proc. Int. Conf. on Acoustics, Speech and Signal Processing*, Vol. 1 pages 133–136, 1993.

[26]  A. S. Bregman. *Auditory Scene Analysis*. MIT Press, 1990.

[27]  H. Purwins, B. Blankertz, and K. Obermayer. Computing auditory perception. *Organised Sound*, **5**(3): 159–171, 2000.

# 5 Speech communications

Chapters 1, 2 and 3 described the foundations of speech signal processing – the characteristics of audio signals in general, methods of handling and processing them – and the features of speech as produced and understood by humans. In particular we have covered some basic MATLAB methods for handling speech and audio which we will build upon in this chapter as we embark upon an exploration of the handling of speech signals in more depth.

This chapter will consider typical speech handling in terms of speech coding and compression (rather than in terms of speech classification and recognition, which often use similar techniques but are higher level in nature). We will first consider quantisation of speech, which assumes that speech is simply a general audio waveform (i.e. it does not incorporate any knowledge of the characteristics of speech).

Knowledge of speech features and characteristics allows for parameterisation of the speech signal, and then source filter modelling which will be considered in turn. Perhaps the pinnacle of achievement in these approaches is the CELP (Codebook Excited Linear Prediction) speech compression techniques, which will be discussed in the final section.

---

**Infobox 5.1** Speech coding objectives

Speech compression, or codec systems, are classified according to what they compress: speech, or general audio, how well they compress this, and how well they perform in terms of quality or intelligibility (which were differentiated and measured in Section 3.3.1). To aid in this classification, there is a general agreement on terms used to describe the quality of speech handled by each method. The table below lists the more common terms, and describes them in terms of sample rate, bandwidth, approximate dynamic range and mean opinion score (MOS – see Section 3.3.2). All figures given are approximate guides to the typical characteristics of such systems:

| Name | Sample rate | Bandwidth | Dynamic range | MOS |
|------|-------------|-----------|---------------|-----|
| synthetic quality | – | – | 48 dB | 2.5–3.5 |
| communications quality | 7200 Hz | 200–2000 Hz | 56 dB | 3.5–4.0 |
| toll quality | 8000 Hz | 200–3200 Hz | 64 dB | 4.0 |
| network quality | 16 000 Hz | 20–7000 Hz | 80 dB | 4.0–4.5 |

Toll quality refers to 'telephone audio', based on the analogue telephone network, but often brought into the realm of digital measurements. For analogue systems a signal-to-noise ratio of 30 dB, and 200 Hz to 3.2 kHz bandwidth, measured at the 3 dB points, is typical.

---

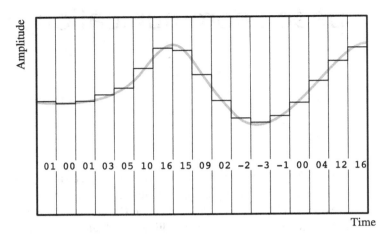

01  00  01  03  05  10  16  15  09  02  −2  −3  −1  00  04  12  16

Time

**Figure 5.1** Illustration of an audio waveform being pulse code modulated by quantising the amplitude of the signal to the nearest integer at sampling points spaced regularly in time.

## 5.1    Quantisation

As mentioned at the beginning of Chapter 1, audio samples need to be quantised in some way during the conversion from analogue quantities to representations on computer. In effect, the quantisation process is reducing the amount of information stored: the fewer bits of quantisation, the less audio information is captured.

Most real-world systems are bandwidth (rate) or size constrained, such as an MP3 player only being able to store 1 Gbyte of audio. Obviously to get the most out of such a device it is necessary to reduce the number of bits required to store the audio, but without compromising quality too much. This is generally known as audio compression. A large part of handling speech in communication systems is in determining how to reduce the number of bits stored or conveyed, whilst maximising quality or intelligibility.

### 5.1.1    Pulse coded modulation

Pulse coded modulation (PCM) is the format delivered by most analogue-to-digital converters (ADCs) and the format of choice for representing audio on a computer. The sound is stored as a vector of samples, with each sample usually (but not always) represented as a single 16-bit value. The samples are supposed to be related to the analogue amplitude of the audio waves travelling through the air in some way, with the timing between samples being determined by the sample rate. This is shown in Figure 5.1 where a waveform is time-sliced, and the average amplitude in each time slice encoded along the bottom. These values form the sample vector describing the waveform.

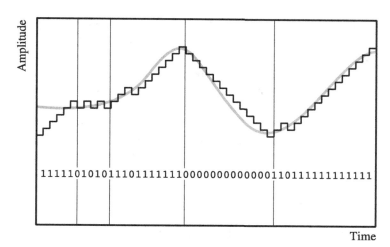

Time

**Figure 5.2** Illustration of an audio waveform being represented using delta modulation, where a 1 indicates a stepwise increase in amplitude and a 0 indicates a stepwise decrease in amplitude, at each sampling position.

Most discrete digital signal processing relies upon PCM samples being available, however there has recently been a resurgence of interest in non-uniform sampling representations – these are vectors that have a higher sampling frequency around regions of the signal representing important information, or perhaps containing higher frequency components. Perhaps in time such techniques will find their way into mainstream algorithms, but for now the difficulty in understanding the theory of such systems, and working around the practical difficulties involved in their execution, have limited their usefulness.

## 5.1.2    Delta modulation

The 'delta' refers to a difference value from one instant to the next, with the difference being limited to $+1$ or $-1$. This system maintains an accumulator that starts at zero. At every sample position, this accumulator either steps up by one or steps down by one, never maintaining the same value. Stepping up or down is in response to a comparison between the current accumulator value with the desired waveform amplitude at the same time (and the decision is 'are we too low? then step up, otherwise step down'). This is shown in Figure 5.2, and it should be noted that this method would typically have a far higher sample rate than PCM audio – but then only require a 1 or 0 (up or down) at each sample instant.

The problem with delta modulation is that the quantisation error depends on the step-size. This means that in order to represent audio as accurately as possible, the step height should be small. However a small step means that more steps are needed to reach up to larger waveform peaks. In Figure 5.2 when rising up to the first peak, and dropping down after it, there is a large gap between the waveform we desire to quantise and the actual step values – this is because 'delta mod' can only increase a single step at a time, but the gradient of the waveform has exceeded this. Such a limit on the gradient is termed the *slew rate* or sometimes *slope overload*. The output bitstream, representing the waveform in the delta modulation format, is shown below the waveform.

In order to support a higher slew rate without increasing the stepsize, it is necessary to sample more quickly. This is a trade-off between bits per second used in the representation and the quantisation noise introduced by it.

### 5.1.3    Adaptive delta modulation

In an attempt to maintain the beneficial features (and simplicity) of delta modulation, but to overcome the slew rate limitations, designers came up with several methods to vary the step height based on the past history of the audio. These are adapting the quantisation level so it can be small when dealing with slowly changing waveforms, but coarse when dealing with rapid waveform changes. The technique is also known by the mouthful *continuously variable slope delta modulation*, abbreviated CVSDM or CSVD, and used as a speech compression method in some older radio systems.

In the most basic form, such a system relies upon some rules to change stepsize, such as the following artificial example:

'If the past $n$ values were the same then double the step height, otherwise halve it.'

There would, of course, be upper and lower limits to the step height changes. In reality, some systems would themselves gradually adapt their stepsize rules over time. Often step heights gradually increased, and gradually decreased (rather than the system mentioned which doubles it or halves it each time, considered a fairly large change).

The technique is illustrated in Figure 5.3, for $n = 3$, and thus the step height can be seen to change following three successive moves in the same direction. Several step height reductions are similarly illustrated. The bitstream resulting from this process, which is a quantised representation of the original waveform, is shown across the bottom of the plot.

### 5.1.4    ADPCM

Adaptive differential pulse coded modulation (ADPCM) is a method of applying the adaptive delta modulation technique to PCM samples (instead of to a delta-modulated bitstream). The word 'differential' indicates that the system calculates a difference at each sample instant, and 'adaptive' indicates that stepsizes change based on past history.

Similar to delta modulation, there is an accumulator that starts at zero. At the first sample instant a difference is calculated between the accumulator and the waveform being coded. This difference is quantised and then added to the accumulator. The next waveform sample value is then compared to the new accumulator value, and a difference

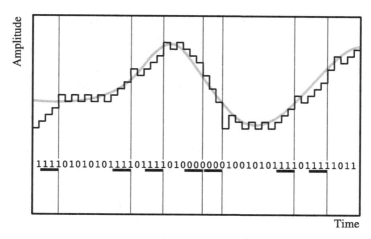

**Figure 5.3** Illustration of an audio waveform being represented using adaptive delta modulation, where a 1 indicates a stepwise increase in signal amplitude and a 0 indicates a stepwise decrease in signal amplitude. Repetition of three like sample values triggers a doubling of stepsize, and a halving of stepsize is triggered by the neighbouring samples which are unlike. The predominant stepsize in the figure is the minimum limit, and there would likewise be a maximum stepsize limit imposed.

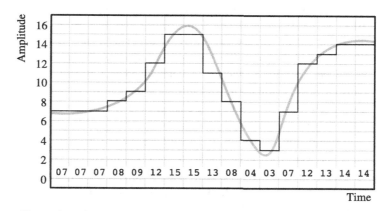

**Figure 5.4** Illustration of an audio waveform being quantised to 16 levels of PCM.

calculated, quantised and then added to the accumulator. Evidently the difference can be either positive or negative, and it is thus the quantised difference value that is transmitted at each sample instant.

The adaptive nature of the system comes into play during the quantisation stage, by adjusting the size of the quantisation steps. Typically, a 16-bit sample vector would be encoded to 3-, 4- or 5-bit ADPCM values, one per sample.

As an example, consider the artificial PCM sample stream in Figure 5.4, with each sample being quantised to 16 levels (in reality a sample stream would be quantised to 65 536 levels in a 16-bit system, but for the sake of clarity the illustration is simplified). At each sampled time instant the value of the quantised level represents the waveform amplitude. In this case the PCM sample vector would be {07 07 07 08 09 12 15 15 13 08 04 03 07 12 13 14 14}.

If we were to use a differential PCM scheme, then we would calculate the difference between each sample. Assume we start with an accumulator of zero, the differences for the sample vector shown would be {07 00 00 01 01 03 03 00 −02 −05 −04 −01 04 05 01 01 00}. Apart from the first sample, we can see that the differential vector values are much smaller (i.e. we could use fewer bits to represent them).

ADPCM uses this methodology but takes it one step further by changing the quantisation stepsize at each sampling instant based on past history. For an example, the same waveform is coded, using the same number of bits per sample in Figure 5.5. Starting with the same initial quantisation levels, the rule used here is that if the sample value is in the middle four quantisation levels then, for the next sample, the quantisation stepsize is halved, otherwise it is doubled. This allows the quantisation to zoom in to areas of the waveform where only small sample changes occur, but to zoom out sufficiently to capture large changes in amplitude also.

In this case the adaptively quantised PCM sample vector would be {07 08 09 10 10 11 11 09 07 05 04 05 14 12 08 10 08} and once we have used differential coding on this it would become {07 01 01 01 00 01 01 −02 −02 −02 −02 01 09 −02 −04 02 −02}. The diagram shows how the quantisation step zooms in on slowly changing waveform amplitudes, and then zooms out to capture large changes in amplitude. This reduces quantisation noise, and still manages to achieve high slew rate (except where a flat waveform is followed by big changes and conversely when large waveform features are followed immediately by much smaller ones).

In reality, the adaptation rule would use the last few samples rather than just the last one – and would be used to predict the next waveform value at the encoder, then compare the predicted value to the actual value. ADPCM is actually quite good at tracking a waveform – especially where the waveform values can be reasonably accurately predicted such as in harmonic music.

## 5.1.5    SB-ADPCM

The SB-ADPCM coder, introduced through ITU standard G.722, includes two separate ADPCM coding units. Each one operates on half of the frequency range (which are 0–4 kHz and 4–8 kHz respectively), but the bit weighting and quantisation differ by a factor of 4 in favour of the lower band, which is thus capable of better audio fidelity, particularly because it conveys frequencies that are more important to speech (see Section 3.2.4).

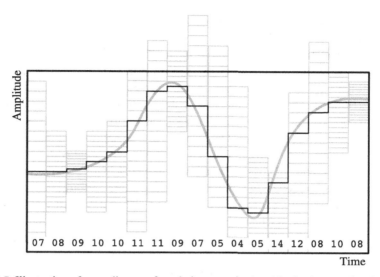

**Figure 5.5** Illustration of an audio waveform being quantised to 16 adaptive step levels. The figure shows that both the absolute placement and the size of the quantisation steps is determined dynamically at each sample point based upon the previous quantised sample values.

Remember the masking properties of the human auditory system (Section 4.2.8)? This states that loud tones will mask nearby, but quieter frequencies. Similarly ADPCM tends to match the loudest sounds by varying its quantisation step to match them, leaving quieter sounds to be lost in the quantisation noise. Since this follows in many ways the characteristics of the human auditory system, it is not a great disadvantage – except when the quiet sound is far away in frequency from the loud sound. In humans such a sound would no longer be masked (since it would now be in a different critical band – see Section 4.3.2), but ADPCM has no concept of frequency bands and would probably lose the sound. For this reason, SB-ADPCM, being able to simultaneously code one very loud and one very quiet sound in different frequency ranges, is perceived as having much higher quality than the ADPCM equivalent.

## 5.2     Parameterisation

Coding techniques that follow, or try to predict, a waveform shape tend to be relatively simple and consequently achieve limited results. These techniques typically assume very little about the waveform being coded – except perhaps maximum extents and slew rate. There is a trade-off between coding quality and bitrate, and very little room to manoeuvre toward the ideal of a high-fidelity coding scheme with very low bitrate.

Instead of coding the physical waveform directly, researchers hit upon the idea of parameterising the sound in some way: several values are chosen to represent important aspects of the speech signal. Whatever parameters are chosen to represent the waveform are then transmitted from coder to decoder, where they are used to recreate a similar (but not identical) waveform.

Apart from the likelihood of the transmitted parameters requiring fewer bits to represent than a directly coded waveform, parameterisation can hold two other benefits. Firstly if the parameters are chosen to be particularly relevant to the underlying sound (i.e. a better match to the speech signal) then the difference between the original and coded-decoded signal can be reduced, leading to better fidelity. Second is that the method of quantising the parameters themselves – or rather the number of bits assigned to each parameter – can be carefully chosen to improve quality. In more simple terms, when given a 'pool' of bits that are allowed to represent the parameters being transmitted from encoder to decoder, it is possible to 'spend' more bits on parameters that are more important to overall quality (where the measure of quality itself will be considered alongside intelligibility in Chapter 6). In the more advanced speech coding algorithms, parameters are chosen that match the component signals within the speech, or that match the important aspects of the human auditory system, or perhaps cater a little to both speech production and speech understanding.

Figure 5.6 shows the process used in a great many modern speech coding systems, where the original speech signal is split into components that describe the overall gain or amplitude of the speech vector being coded, the pitch information, vocal tract resonances, and lung excitation. Each of these parameters is to be quantised and transmitted from encoder to decoder. At the decoder the parameters are used together to recreate synthesised speech.

---

**Infobox 5.2** PCM-based speech coding standards

As you might expect there are a plethora of different speech coding techniques based around PCM. The main standardisation body in this area is the International Telecommunications Union (ITU) since the primary driver in voice communications worldwide has been telecommunications. In the past the ITU was known as the Comité consultatif international téléphonique et télégraphique or CCITT. Even in English the name is a mouthful: 'International Telegraph and Telephone Consultative Committee', hence the move to the far simpler title ITU.

Several of the more common ITU standards, all beginning with the prefix G are shown in the following table:

| Name | Description |
|------|-------------|
| G.711 | 8 kHz sampling A-law and $\mu$-law compression |
| G.721 | 32 kbits/s ADPCM standard (replaced by G.726) |
| G.723 | 24 and 40 kbits/s ADPCM (replaced by G.726) |
| G.722 | 64 kbits/s SB-ADPCM sampled at 16 kHz |
| G.726 | 24, 32 and 40 kbits/s ADPCM sampled at 8 kHz |

Several other ITU speech standards are shown on page 131.

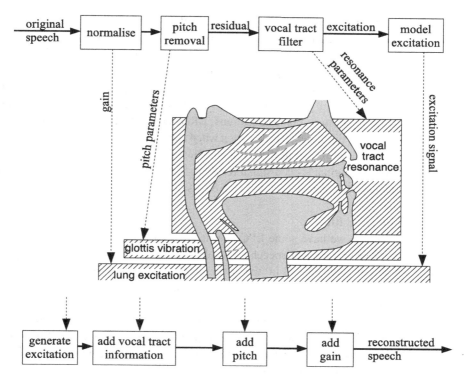

**Figure 5.6** Parameterisation of the speech signal into components based loosely upon the human speech production system.

In the sections that follow, many speech coding techniques will be introduced and discussed – but all related back to this parameterisation of the speech signal into different components that relate in some way to how they were produced.

## 5.2.1 Linear prediction

Linear prediction has, for several decades, been the mainstay of speech communications technology, and applied to speech coding since at least 1971 [1]. It relies upon several characteristics of speech derived from the fact that speech is produced by a human muscular system. These muscles act to shape speech sounds through their movement, which is limited by a maximum speed. Muscles cannot move infinitely fast and thus, although human speech varies considerably throughout an utterance, it actually remains pseudo-stationary for around 30 ms (explained further in Section 2.5.1).

In actual fact, the action of the glottis to generate pitch spikes is often shorter than 30 ms, so through some clever processing (see Section 5.3.2) pitch needs to be removed from the speech signal first – leaving a much lower energy signal called the *residual*.

Pseudo-stationarity implies that the 240 samples at 8 kHz sample rate (corresponding to a duration of 30 ms), being similar, can be parameterised by a smaller set of values: typically eight or 10 linear prediction coefficients. Linear prediction coefficients are generator polynomials for a digital filter that, when stimulated with some input signal, recreates the characteristics of the original samples. They may not appear identical in the time domain, but most importantly their frequency response will match the original. Linear predictive coding (LPC) has been used successfully, by itself, in speech coding: the very low bitrate US Federal Standard 1015 2.4 kbits/s algorithm, developed in 1975, is based on LPC. It does, however, have very low perceived quality, and therefore limited to use in military communications. More normally LPC is refined with some other techniques discussed in this section when used in more modern coding algorithms. In this section we will cover the LPC filter itself, and then look at how to extract LPC parameters from the original speech signal residual. Some other important features relating to LPC stability and quantisation will also be discussed.

### 5.2.1.1     The LPC filter

Assuming we have some LPC coefficients that describe the vocal characteristics of a 30 ms or so vector of speech residual, then these can be used in two ways. Firstly in a synthesis filter to 'add in' the vocal characteristics to a sample vector. Secondly in an analysis filter to remove the characteristics.

For a $P$th-order linear prediction filter represented by $P$ coefficients $a[0], a[1] \ldots a[P-1]$, the LPC synthesis filter is an all-pole infinite impulse response (IIR) filter, shown for current sample $n$ in Equation (5.1). $x(n)$ is the input audio vector, and $y(n)$ is the vector of output audio, which would have the vocal characteristics encoded in $a$ added in to it:

$$ y(n) = x(n) + \sum_{p=0}^{P-1} a(p)y(n-p). \tag{5.1} $$

In MATLAB, the linear prediction coefficients are easy to generate using the `lpc()` function, which uses the Durbin–Levinson–Itakura method (which we will explore later in Section 5.2.2) to solve the recursive equations necessary to generate LPC coefficients.

Assume we have a MATLAB recording of speech in some vector which is imaginatively named `speech`, we would normally analyse this one pseudo-stationary window at a time, using overlapping windows. For 8 kHz sampling rate, and pseudo-stationarity of 20–30 ms, we would thus define an analysis window of 160–240 samples (see Section 2.5.1). For illustrative purposes, we will manually segment a speech recording, window it, and perform LPC analysis to obtain a tenth-order set of linear prediction coefficients:

```
seg=speech(1:160);
wseg=seg.*hamming(160);
a=lpc(wseg,10);
```

For the sake of demonstration and comparison, here is an example set of LPC coefficients which we can use for testing:

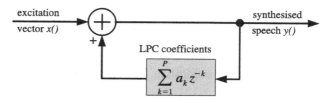

**Figure 5.7** Use of LPC coefficients in a synthesis filter.

```
a=[1; -1.6187;2.3179; -2.9555;2.8862; -2.5331;2.2299;
-1.3271;0.9886; -0.6126;0.2354];
```

Note that the set begins with a 1.0, and this is standard for LPC coefficient sets within MATLAB. However when transmitting or storing LPC coefficients of speech, we may delete the leading 1.0 before storage/transmission, always remembering to add it in again during recall/reception.

In $z$-transform terms, the LPC coefficients are denoted $A(z)$ such that the synthesis filter $H(z)$ is:

$$H(z) = \frac{1}{A(z)} = 1/\left\{1 + \sum_{i=1}^{P} a_i z^{-i}\right\}. \tag{5.2}$$

Note the '1+' in the final denominator which leads to a block diagram structure as shown in Figure 5.7, implementing the direct form of the expression in Equations (5.1) and (5.2), and which mirrors the leading 1.0 of the LPC parameter set in MATLAB.

The LPC analysis filter shown in Figure 5.8, acts to remove the vocal tract information from a signal, but since this is a digital filter, it has a spectral response which we can analyse or plot. In general, we would do this by substituting $z^{-1}$ in Equation (5.2) with a sweep from $z = 0$ up to $z = e^{-j\omega}$, effectively sweeping from DC up to the Nyquist frequency.

In MATLAB we can cheat by using the `freqz()` function to do this and plot the magnitude and phase response. A synthesis filter using the LPC coefficients would create the formants and other resonances as seen in the vocal tract, and can be plotted using:

```
freqz(1, a);
```

On the other hand an analysis filter using these same coefficients would act to remove the resonance from a signal:

```
freqz(a);
```

The `freqz` function not only plots a handy magnitude and phase response graph, but can also return these as complex arrays which we can use for subsequent analysis

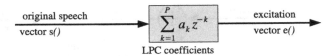

**Figure 5.8** Use of LPC coefficients in an analysis filter.

algorithms. As an example we can obtain the magnitude response of the synthesis filter at $N$ equally-spaced frequencies, and then determine the maximum resonance peak (perhaps corresponding to formant F1 in a speech sample), assuming an 8 kHz sample rate as follows:

```
Fs=8000; %sample rate
N=100; %frequency resolution
[H, F] = freqz(1,a,N);
%Plot the magnitude with a logarithmic scale on the y-axis
semilogy(0:Fs/(N-1):Fs,abs(H));
[y,n]=max(abs(H));
PeakF=(n-1)*Fs/(N-1);
```

The peak frequency of the plot is returned, along with the array index, by the max() function, and because MATLAB indexes arrays from element 1, it is necessary to adjust the returned index, $n$, when determining the value in Hz.

Both analysis and synthesis forms are used within LPC-based speech compression algorithms – one (or more) at either end. This is indicated in Figure 5.9 which shows a generic LPC-based speech coding algorithm. The same diagram, with some differences in how the excitation is derived or managed, probably represents all CELP class coders (which we will see later in Section 5.4.1), and also the GSM-style pulse-excitation coders (Section 5.3.1).

In the encoder, original speech is normalised to be at some predetermined amplitude range, then the pitch information extracted to produce a residual. The residual, as mentioned previously, contains vocal tract information which is modelled by LPC. Applying the LPC analysis filter to the residual will result in the vocal tract information being (mostly) removed, leaving a lung excitation signal which is modelled in some way and then transmitted to the decoder.

The decoding process is essentially the reverse of the encoding process, and results in reconstructed speech – the fidelity of which depends on the ability of the algorithm to model the signals being parameterised, and the degree of quantisation applied to the parameters passed from encoder to decoder. Note that in the diagram of Figure 5.9 the extraction process (by which the various parameters are determined), has not been shown. Each of the parameters conveyed by the coder will be discussed in the sections which follow.

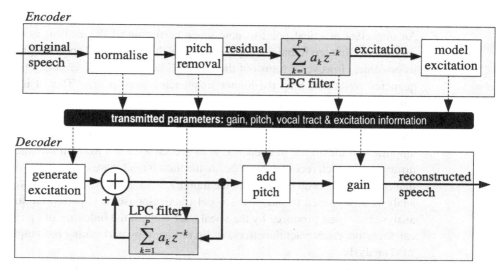

**Figure 5.9** The role of LPC analysis and synthesis filters in many speech coding algorithms to convey parameterised vocal tract information from encoder to decoder.

### 5.2.1.2    LPC stability issues

Typically the LPC parameters $a(\ )$ are represented in MATLAB as floating point values. If these were used in a speech coder to represent speech parameters then they would need to be saved as smaller fixed-point values for transmission between encoder and decoder (since floating point values occupy too many bits).

Unfortunately this quantisation process can occasionally produce a set of parameters that results in an unstable synthesis filter in the decoder. This means that the LPC filter output magnitude rises sharply then 'blips' towards infinity before recovering. When listening to speech this effect would be recognisable as a loud pop or squeak. The same effect can sometimes be heard on live news broadcasts from remote locations by satellite phone – pops and blips in the audio disrupting a reporter's voice are the result of errors in the parameterised audio bitstream 'kicking' the LPC synthesis filter into momentary instability.

Of course, in the presence of bit errors in the parameters such things are expected, however even in the absence of errors, directly quantising LPC parameters often results in instability. We will look briefly at the effect of quantising LPC parameters directly in Section 5.2.5, but note that LPC parameters are never quantised directly – in practice they are always converted into an alternative form. One such form is the reflection coefficients described in Section 5.2.2. Log Area Ratios (LAR) are the method of choice for the GSM speech coder, comprising a logarithmic transformation of the reflection coefficients, and other more esoteric mathematical transformations exist. None of these can compete with the quantisation characteristics of Line Spectral Pairs (LSPs) which are worthy of their own discussion in Section 5.2.4.

### 5.2.1.3    Pre-emphasis of the speech signal

An important practical point to note when performing LPC analysis is that the LPC coefficients that are found are supposed to match the analysed speech signal as closely as possible. However it turns out that the LPC equations tend to satisfy the lower frequencies while matching the higher frequencies more poorly. Thus it is common to emphasise the higher frequencies prior to LPC analysis.

In fact, when speech is radiated from a human mouth, from an area of high pressure, through a constriction, into a low pressure area, a spectral roll-off occurs to reduce the amplitude of the higher frequencies. Thus speech recorded from *outside* the mouth will differ from speech recorded from *inside* the mouth (and there do exist tiny microphones that can record speech from within the mouth). Since the LPC filter employed in speech analysis is supposed to model the vocal tract response, it is preferable to allow it to analyse the signal produced by the vocal tract without the influence of lip radiation. We can therefore counteract the effects of lip radiation by performing pre-emphasis before LPC analysis.

Given a speech signal $s(n)$, the pre-emphasis would normally be performed with a single-tap filter having transfer function $(1 - \alpha z^{-1})$, and an emphasis coefficient, $\alpha$ of nearly 1. A typical value used in research systems is $\alpha = 15/16 = 0.9375$. Thus each pre-emphasised speech sample $s'(n)$ comes from the current and previous input speech samples acted on by the following FIR filter:

$$s'(n) = s(n) - 0.9375 \times s(n-1). \tag{5.3}$$

By performing pre-emphasis of the speech signal in this way prior to LPC analysis, we can better approach the signal that leaves the vocal tract, and can overcome one of the issues with LPC analysis where the coefficients match the higher frequency components poorly.

Of course, any speech system that outputs something based upon pre-emphasised speech will sound a little strange – we do not normally hear speech from *within* a person's mouth. So even though the processing can be conducted on pre-emphasised speech, the output must be de-emphasised, to replace the attenuated low frequencies which we had removed.

The de-emphasis filter matches the emphasis filter, reversing the emphasis applied to the speech signal $s'(n)$ to recreate more natural sounding speech $r(n)$. This IIR filter is as follows:

$$r(n) = s'(n) + 0.9375 \times r(n-1). \tag{5.4}$$

In MATLAB we can create a pre-emphasis filter very easily, and apply this using the filter() function. If the original speech signal is called s, the pre-emphasised output is to be es and the de-emphasised version of this is ds then we can easily convert between them:

```
% Create the emphasis/de-emphasis filter coefficients
```

```
h=[1, -0.9375];
% Apply the emphasis filter
es=filter(h, 1, s);
% Apply the de-emphasis filter
ds=filter(1, h, es);
```

It is instructive to try listening to this on a piece of test speech and to use the MATLAB `psd()` function to conveniently plot a frequency response of the signals at each stage. The slightly nasal sounding es is a sound that many speech researchers are familiar with. If, when developing a speech algorithm, you hear something like this, then you will realise that you have lost some of the lower frequencies somewhere, or perhaps forgotten the de-emphasis filter.

### 5.2.2    Reflection coefficients

Since the LPC coefficients by themselves cannot be reliably quantised without causing instability, significant research effort has gone into deriving more stable transformations of the parameters. The first major form are called *reflection coefficients* because they represent a model of the synthesis filter that, in physical terms, is a set of joined tubes of equal length but different diameter.

In fact the same model will be used for Line Spectral Pairs (Section 5.2.4) but under slightly different conditions. The reflection coefficients quantify the energy reflected back from each join in the system. They are sometimes called partial correlation coefficients or PARCOR after their method of derivation.

Conversion between PARCOR and LPC is trivial, and in fact LPC coefficients are typically derived from input speech by way of a PARCOR analysis (although there are other methods). This method, and the rationale behind it, will be presented here, and can be followed in either [1] or [2]. First, we make the assumption that, given a frame of pseudo-stationary speech residual, the next sample at time instant $n$ can be represented by a linear combination of the past $P$ samples. This linear combination is given by Equation (5.5):

$$x'[n] = a_1x[n-1] + a_2x[n-2] + a_3x[n-3] + \cdots + a_Px[n-P]. \qquad (5.5)$$

The error between the predicted sample and the actual next sample quantifies the ability of the system to predict accurately, and as such we need to minimise this:

$$e[n] = x[n] - x'[n]. \qquad (5.6)$$

The optimum would be to minimise the mean-squared error over all $n$ samples:

$$E = \sum_n e^2[n] = \sum_n \left\{ x[n] - \sum_{k=1}^{P} a_k x[n-k] \right\}^2. \qquad (5.7)$$

In order to determine the set of LPC coefficients resulting in the minimum mean-squared error, $E$ in Equation (5.7), we must differentiate the expression and then equate to zero:

$$\frac{\delta E}{\delta a_j} = -2 \sum_n x[n-j] \left\{ x[n] - \sum_{k=1}^{P} a_k x[n-k] \right\} = 0. \qquad (5.8)$$

This results in a set of linear equations containing $P$ unknowns, $a$, that can be derived from the known speech samples $x[n]$ to $x[n-P]$:

$$\sum_{k=1}^{P} a_k \sum_n x[n-j]x[n-k] = \sum_n x[n]x[n-j], \qquad (5.9)$$

where $j = 1, \ldots, P$.

There is of course a choice of methods to solve such a set of equations. Most common are the covariance and autocorrelation methods. The former splits the speech into rectangular windowed frames and minimises the error over each frame of $N$ samples. The latter assumes that the signal is stationary with finite energy, with an infinite summation range (which is acceptable if we window the speech before analysis). Covariance tends to be more accurate for small speech frames, but with the sharp cutoffs of a rectangular window, can lead to instability. For this reason, most speech coding algorithms opt for the autocorrelation approach (and hence derive partial correlation coefficients, known as PARCOR, along the way).

For the infinite summation, we first note that the following relationships hold:

$$\sum_{n=-\infty}^{\infty} x[n-j]x[n-k] \equiv \sum_{n=-\infty}^{\infty} x[n-j+1]x[n-k+1] \qquad (5.10)$$

and

$$\sum_{n=-\infty}^{\infty} x[n-j+1]x[n-k+1] \equiv \sum_{n=-\infty}^{\infty} x[n]x[n+j-k]. \qquad (5.11)$$

Using these equivalence relationships, we re-formulate Equation (5.9) as follows:

$$\sum_{k=1}^{P} a_k \sum_{n=-\infty}^{\infty} x[n]x[n+j-k] = \sum_{n=-\infty}^{\infty} x[n]x[n-j]. \qquad (5.12)$$

Note the similarity between this and the standard autocorrelation function in Equation (5.13), where $R(k)$ denotes the $k$th autocorrelation:

$$R(k) = \sum_{n=-\infty}^{\infty} x[n]x[n+k]. \qquad (5.13)$$

The set of $P$ linear equations can thus be represented as the following matrix:

$$
\begin{bmatrix}
R(0) & R(1) & R(2) & \dots & R(P-1) \\
R(1) & R(0) & R(1) & \dots & R(P-2) \\
R(2) & R(1) & R(0) & \dots & R(P-3) \\
\vdots & \vdots & \vdots & \ddots & \vdots \\
R(P-1) & R(P-2) & R(P-3) & \dots & R(0)
\end{bmatrix}
\begin{bmatrix}
a_1 \\ a_2 \\ a_3 \\ \vdots \\ a_P
\end{bmatrix}
=
\begin{bmatrix}
R(1) \\ R(2) \\ R(3) \\ \vdots \\ R(P)
\end{bmatrix}. \tag{5.14}
$$

In practice a window, usually Hamming (Section 2.4.2), is applied to the input speech prior to calculating the autocorrelation functions, and the entire autocorrelation results are usually normalised by dividing by $R(0)$ first. These normalised coefficients are denoted $r(i)$.

Standard techniques exist for the matrix solution including brute force matrix inversion, the famous Durbin–Levinson–Itakura method, or the Le Roux method which is slightly less efficient but is a compact and easily followed recursive formula [3]:

$$
k_{n+1} = \frac{e^n_{n+1}}{e^n_0} \quad \text{for} \quad n = 0, \dots, P \tag{5.15}
$$

$$
e^{n+1}_0 = e^n_0 - k_{n+1} e^n_{n+1} = e^n_0 (1 - k^2_{n+1}) \tag{5.16}
$$

$$
e^{n+1}_i = e^n_i - k_{n+1} e^n_{n+1-i} \quad \text{for} \quad i = n, \dots, P \tag{5.17}
$$

where the initial conditions for the recursion are set to $e^0_i = R(i)$ for each $i$ in the set of $P$ equations.

The values of $k$ that are obtained from the Le Roux method are the *reflection coefficients*.

## 5.2.3 Converting between reflection coefficients and LPCs

Conversion from the reflection coefficients (which in some cases are quantised and transmitted from speech encoder to decoder) to the LPC parameters which are required for the LPC synthesis filter to recreate encoded speech, is not difficult. The relationship is shown in Equation (5.18) for all $P$ coefficients, where the notation $a^i_j$ indicates the $j$th LPC coefficient at time instant $i$:

$$
a^i_j = a^{(i-1)}_j + k_i a^{(i-1)}_{(i-j)} \quad \text{with} \quad 1 \le j \le i - 1. \tag{5.18}
$$

In order to perform the reverse conversion from LPC parameters into reflection coefficients, we start with the initial value:

$$
k_i = a^i_j
$$

and then follow with:

$$a_j^{(i-1)} = \frac{a_j^i - a_i^i a_{(i-j)}^i}{1 - k_i^2} \quad \text{with} \quad 1 \leq j \leq i-1 \tag{5.19}$$

where Equation (5.19) is repeated with $i$ decreasing from $P$ to 1 with initial conditions of $a_j^P = a_j$ for all $j$'s between 1 and $P$.

### 5.2.4     Line spectral pairs

Line spectral pairs (LSPs) are a direct mathematical transformation of the set of LPC parameters, and are generated within many speech compression systems, such as the more modern CELP coders (which will be discussed later in Section 5.4.1). LSP usage is popular due to their excellent quantisation characteristics and consequent efficiency of representation. They are also commonly referred to as Line Spectral Frequencies (LSF) [3].

LSPs collectively describe the two resonance conditions arising from an interconnected tube model of the human vocal tract. This includes mouth shape and nasal cavity, and forms the basis of the underlying physical relevance of the linear prediction representation. The two resonance conditions describe the modelled vocal tract as being either fully open or fully closed at the glottis respectively (compare this to the model of the reflection coefficients in Section 5.2.2). The model in question is constructed from a set of equal length but different diameter tubes, so the two conditions mean the source end is either closed or open respectively. The two conditions give rise to two sets of resonant frequencies, with the number of resonances in each set being determined by the number of joined tubes (which is defined by the order of the analysis system). The resonances of each condition give rise to odd and even line spectral frequencies respectively, and are interleaved into a set of LSPs which have monotonically increasing value.

In reality, however, the human glottis opens and closes rapidly during speech – it is neither fully closed nor fully open. Hence actual resonances occur at frequencies located somewhere between the two extremes of each LSP condition. Nevertheless, this relationship between resonance and LSP position lends a significant physical basis to the representation. Figure 5.10 illustrates LSPs overlaid on an LPC spectral plot (made using the `lpcsp()` MATLAB function given later in Section 5.2.4.3). The 10 vertical lines were drawn at the LSP frequencies, and show the odd and even frequencies being interleaved. Both the LSPs and the spectrum were derived from the same set of tenth-order linear prediction parameters which were obtained from a linear predictive analysis of a 20 ms voiced speech frame.

Notable features of Figure 5.10 include the natural interleaving of the odd and even LSP frequencies, and the fact that spectral peaks tend to be bracketed by a narrow pair of lines (explained by the comment previously indicating that the actual resonance

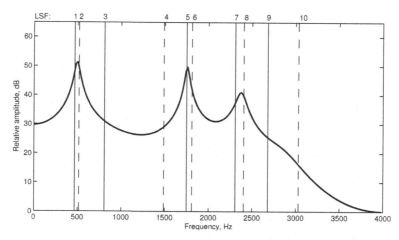

**Figure 5.10** Plot of a sample LPC spectrum with the corresponding LSP positions overlaid. Odd lines are drawn solid and even lines are drawn dashed.

frequency lies somewhere between the open and closed model positions, represented by odd and even lines). Local minima in the spectrum tend, by contrast, to be avoided by nearby lines. These and several other properties explain the popularity of LSPs for the analysis, classification and transmission of speech.

### 5.2.4.1 Derivation of LSPs

Line spectral frequencies are derived from the linear predictive coding (LPC) filter representing vocal tract resonances in analysed speech as we have seen in Section 5.2.1.1. For $P$th-order analysis, the LPC polynomial would be:

$$A_p(z) = 1 + a_1 z^{-1} + a_2 z^{-2} + \cdots + a_P z^{-P}. \tag{5.20}$$

We will define two $(P + 1)$th-order polynomials related to $A_p(z)$, named $P(z)$ and $Q(z)$. These are referred to as antisymmetric (or inverse symmetric) and symmetric in turn based on observation of their coefficients. The polynomials represent the interconnected tube model of the human vocal tract and correspond respectively to complete closure at the source end of the interconnected tubes and a complete opening, depending on an extra term which extends the $P$th-order polynomial to $(P + 1)$th-order. In the original model, the source end is the glottis, and is neither fully open nor fully closed during the period of analysis, and thus the actual resonance conditions encoded in $A_p(z)$ are a linear combination of the two boundaries. In fact this is simply stated:

$$A_p(z) = \frac{P(z) + Q(z)}{2}. \tag{5.21}$$

The two polynomials are created from the LPC polynomial with an extra feedback term being positive to model energy reflection at a completely closed glottis, and negative to model energy reflection at a completely open glottis:

$$P(z) = A_p(z) - z^{-(P+1)}A_p(z^{-1}) \tag{5.22}$$

$$Q(z) = A_p(z) + z^{-(P+1)}A_p(z^{-1}). \tag{5.23}$$

The roots of these two polynomials are the set of line spectral frequencies, $\omega_k$. These relate to symmetric and antisymmetric polynomials as will be shown later in Equation (5.25).

The complex roots of the polynomials can be shown to lie on the unit circle in the $z$-plane if the original LPC filter was stable [3], and alternate around the unit circle. Note that *any* equivalent size set of roots that alternate in this way around and on the unit circle will represent a stable LPC filter. This is something with implications which we will exploit for LSP-based speech processing in Chapter 7.

There are many different ways of calculating or estimating roots, including trial and error, Newton–Raphson approximation, evaluating around the unit circle looking for sign changes, and so on, with the exact choice of method having implications for accuracy, processing complexity and processing delay. In MATLAB we can simply use the in-built `roots()` function to perform the calculation.

If we denote the set of complex roots as $\theta_k$, then the line spectral frequencies are determined from solving Equations (5.22) and (5.23).

$$\omega_k = \tan^{-1}\left(\frac{\mathrm{Re}\{\theta_k\}}{\mathrm{Im}\{\theta_k\}}\right). \tag{5.24}$$

$\omega_k$ are the $P$ line spectral frequencies expressed in radians.

Despite the seemingly complex equations, an example MATLAB script for converting from LPC to LSP is relatively short:

```
function lsp=lpc_lsp(a)
p=length(a); A(1)=1; B(1)=1;
  for k=2:p
  A(k)=(a(k) - a(p+2-k)) + A(k-1);
  B(k)=(a(k) + a(p+2-k)) - B(k-1);
end
r1=roots(A);
r2=roots(B);
for k=1:p-1
   if (real(r1(k)) < 0)
      theta1(k)=pi-abs(atan(imag(r1(k))/real(r1(k))));
   else
      theta1(k)=abs(atan(imag(r1(k))/real(r1(k))));
   end
   if (real(r2(k)) < 0)
      theta2(k)=pi-abs(atan(imag(r2(k))/real(r2(k))));
```

```
  else
    theta2(k)=abs(atan(imag(r2(k))/real(r2(k))));
  end
end
p=p-1;
for k=1:p/2
  theta(k)=theta1(k*2);
  theta(k+(p/2))=theta2(k*2);
end
lsp=sort(theta);    %Sort into ascending order
```

For testing purposes, an example set of LPC coefficients from tenth-order analysis of speech is given below, with the corresponding LSP converted using the function above (these are the same coefficients used previously for the illustration of LPC coefficients in Section 5.2.1.1):

```
a=[1;-1.6187;2.3179;-2.9555;2.8862;-2.5331;2.2299;-1.3271;
   0.9886;-0.6126;0.2354];
lsp=[0.3644;0.4023;0.6334;1.1674;1.3725;1.4205;1.8111;
     1.8876;2.1032;2.3801];
```

Note that, by convention the LPC coefficient array always begins with a fixed '1.0' (also as discussed in Section 5.2.1.1). By contrast, each of the LSP values is meaningful in itself, although as we will examine in Section 5.2.5, they tend to occupy certain frequency ranges most of the time.

### 5.2.4.2 Generation of LPC coefficients from LSPs

Conversion from LSPs to LPCs is a simple process (although the in-depth proof is more difficult – for this refer to the excellent reference book by Saito and Nakata [3]). For this conversion we can, of course, use the ordered LSPs $\omega_k$ to recreate the polynomials that they are roots of, namely $P(z)$ and $Q(z)$ [4]:

$$P(z) = (1 - z^{-1}) \prod_{k=2,4,...,P} (1 - 2z^{-1} \cos \omega_k + z^{-2})$$
$$Q(z) = (1 + z^{-1}) \prod_{k=1,3,...,P-1} (1 - 2z^{-1} \cos \omega_k + z^{-2}) \tag{5.25}$$

and then these can be substituted into Equation (5.21) which expresses $A_p(z)$ as a linear combination of $P(z)$ and $Q(z)$. Since Equation (5.25) involves the cosine of the LSPs, and bearing in mind that some of the more efficient LSP calculation algorithms will yield the roots in the cosine domain [5], it is common also to perform the reverse conversion from LPC to LSP in the cosine domain.

If $q_k$ denotes the array of correctly ordered, lowest-first, LSP frequencies $\omega_k$, in the cosine domain:

$$q_k = \cos \omega_k. \tag{5.26}$$

To perform the conversion, we first need to solve the following set of recursive equations as an intermediate step:

$$
\begin{aligned}
&\text{for} \quad k = 1, \ldots, P \\
&f_k = -2f_{(k-1)}q_{(2k-1)} + 2f_{(k-1)} \\
&\quad \text{for} \quad m = (k-1), \ldots, 1 \\
&\quad f_m = f_m - 2f_{(m-1)}q_{(2k-1)} + f_{(m-2)}.
\end{aligned}
\tag{5.27}
$$

We then apply initial conditions of $f_0 = 1$ and $f_{-1} = 0$, and calculate the coefficients of $g$ similarly as follows:

$$
\begin{aligned}
&\text{for} \quad k = 1, \ldots, P \\
&g_k = -2g_{(k-1)}q_{(2k)} + 2g_{(k-1)} \\
&\quad \text{for} \quad m = (k-1), \ldots, 1 \\
&\quad g_m = g_m - 2g_{(m-1)}q_{(2k)} + g_{(m-2)}.
\end{aligned}
\tag{5.28}
$$

Once the values of $f$ and $g$ have been determined, they form a second set of equations:

$$f'_k = f_k + f_{(k-1)} \qquad \text{for} \quad k = 1, \ldots, P/2 \tag{5.29}$$

$$g'_k = g_k + g_{(k-1)} \qquad \text{for} \quad k = 1, \ldots, P/2 \tag{5.30}$$

which are then averaged to form LPC coefficients from the following:

$$a_k = \tfrac{1}{2}f'_k + \tfrac{1}{2}g'_k \qquad \text{for} \quad k = 1, \ldots, P/2 \tag{5.31}$$

$$a_k = \tfrac{1}{2}f'_{(k-P/2)} - \tfrac{1}{2}g'_{(k-P/2)} \qquad \text{for} \quad k = P/2 + 1, \ldots, P. \tag{5.32}$$

Using MATLAB, we can create a function that reads in LSPs (represented in radians), converts them to cosine domain and then replicates the steps above:

```
function a=lsp-lpc(theta)
p=length(theta);
q=cos(theta(1:p));
f1(10)=1; f1(9)=0;
for i=1:p/2
   f1(10+i)=-2*q(2*i-1)*f1(10+i-1) + 2*f1(10+i-2);
   for k=i-1:-1:1
      f1(10+k)=f1(10+k) - 2*q(2*i-1)*f1(10+k-1) + f1(10+k-2);
   end
end
f2(10)=1; f2(9)=0;
for i=1:p/2
```

```
    f2(10+i)=-2*q(2*i)*f2(10+i-1) + 2*f2(10+i-2);
    for k=i-1:-1:1
        f2(10+k)=f2(10+k) - 2*q(2*i)*f2(10+k-1) + f2(10+k-2);
    end
end
f1b(1)=f1(11)+1;
f2b(1)=f2(11)-1;
for i=2:p/2
    f1b(i) = f1(10+i) + f1(10+i-1);
    f2b(i) = f2(10+i) - f2(10+i-1);
end
for i=1:p/2
    a2(i)     = 0.5*(f1b(i) + f2b(i));
    a2(i+p/2) = 0.5*(f1b((p/2)-i+1) - f2b((p/2)-i+1));
end
a=[1,a2];
```

The example LSP and LPC coefficient values given in Section 5.2.4.1 can again be used for testing of the function.

### 5.2.4.3    Visualisation of line spectral pairs

As discussed previously, line spectral pairs are resident in the frequency domain. Their values can be denominated in Hz, more normally in radians, or as the cosine of their radian value. In this book, unless stated otherwise, all LSP values used in equations and MATLAB code will be given in radians.

Whatever the units, each line is located at a particular frequency. Thus a traditional method of visualising the value of lines is to plot them as an overlay on the LPC spectrum. This was illustrated in Figure 5.10. The MATLAB `freqz()` function generates a frequency response from an LPC polynomial, so we shall use that to plot a spectrum, and then overlay the LSPs on top of this, drawn as lines:

```
function lpcsp(a, lsp)
[HH, FF]=freqz(1, a, 100);
semilogy(abs(HH),'m-');
hold on V=axis;
axis([1,length(FF),V(3),V(4)]);
hold on;
lsc=100/pi;
for lp=1:length(lsp)
        line([1+lsp(lp)*lsc,1+lsp(lp)*lsc], [V(3),V(4)]);
end
hold off;
```

**Table 5.1.** SEGSNR resulting from different degrees of uniform quantisation of LPC and LSP parameters.

| Bits/parameter | LPC | LSP |
|:---:|:---:|:---:|
| 4 | – | −6.26 |
| 5 | −535 | −2.14 |
| 6 | −303 | 1.24 |
| 7 | −6.04 | 8.28 |
| 8 | −10.8 | 15.9 |
| 10 | 19.7 | 20.5 |
| 12 | 22.2 | 22.2 |
| 16 | 22.4 | 22.4 |

## 5.2.5     Quantisation issues

Since LSPs are most often used in speech coders, where they are quantised prior to transmission, it is useful to explore the quantisation properties of the representation. In order to do this, we can use some representative recorded speech, quantise in turn by different amounts, dequantise to recreate speech, and in each case compare the original and dequantised speech.

In fact, this has been done using a large pre-recorded speech database called TIMIT [6]. Several thousand utterances were analysed, LPC and then LSPs derived, and quantised by different amounts. The quantisation scheme used in this case was uniform quantisation, where each parameter is represented by an equal number of bits, ranging in turn, from 4 to 16 bits per parameter.

Tenth-order analysis was used, and a segmental signal-to-noise ratio (SEGSNR, discussed in Section 3.3.2) determined between the original and dequantised speech. Both LSP and LPC quantisation were tested.

Table 5.1 lists the results, where the more positive the SEGSNR value, the more closely the dequantised speech matches the original speech. These results clearly indicate that LPCs are far more susceptible to quantisation effects than are LSPs: down around 5 bits per parameter, the recreated speech resulting from LPC quantisation exhibits sharp spikes of noise and oscillation, totally unlike the original speech. Hence the huge difference between original and recreated speech evidenced by the SEGSNR value of −535. By contrast the LSP representation, with an SEGSNR of −2.14 indicates quite easily understandable speech. This substantiates the assertion in Section 5.2.4 that LSPs are favoured for their superior quantisation characteristics. Finally, note that both approaches in the table achieve a SEGSNR level of 22.4 when more bits are used for quantisation, plainly indicating the limit of achievable SEGSNR for the analysis process used (i.e. the window function, autocorrelation, number of parameters, and so on).

LSPs may be available in either the frequency domain or the cosine domain (depending on the method of solving the polynomial roots). Each line's value can be quantised independently (scalar quantisation) on either a uniform or a non-uniform scale [7] which can also be dynamically adapted. Alternatively, lines can be grouped together and vector quantised with either static or adaptive codebooks [8]. Vector quantisation groups sets of lines together and quantises these as a set. Typical sets may be a $(2, 3, 3, 2)$ or a $(3, 3, 4)$ arrangement for a tenth-order system.

Both scalar and vector quantisation, can be applied either to the raw LSP values themselves, or to differential values, where the difference is either that between a line's current position and its position in the previous frame or between its current position and its mean position [9]. We can refer to these as the short-term and long-term average differences respectively [10].

An adaptation of long-term average differential quantisation (which uses the distance between current position, and mean position of a line), is to recalculate the nominal position every frame based on an even distribution of nominal positions between the values of the first and last LSPs. This is known as Interpolated LSF (or LSF Interpolation, LSFI) [11]. A different form of interpolation is that applied by the TETRA standard CELP coder [12], which quantises LSPs which have been interpolated between subframes (of which there are four per standard-sized frame). This approach can provide a degree of immunity to the effects of subframes lost due to burst errors.

An effective quantisation scheme will generally minimise either the signal-to-quantisation noise ratio for typical signals, or will minimise a more perceptually relevant measure. Such measures could be the commonly-used spectral distortion (SD) value (see Section 3.3.2) or similar variants. Some published examples are the LSP distance (LD) [9], LSP spectral weighted distance measure (LPCW) [13], local spectral approximation weights (LSAW) [14] and inverse harmonic mean weights (IHMW) [15].

In all cases, it is necessary to appreciate the dynamics of the signal to be quantised, and optionally to assign different levels of importance to critical spectral regions, either directly, or by allocating greater quantisation accuracy to LSPs with a frequency locus within such critical regions. It is possible to match regions of spectral importance to LSP accuracy through the selection of different quantiser resolutions for different lines. For example, lines 9 and 10 in a tenth-order analysis would relate to formant F3, if present. This formant can be considered less important to speech intelligibility than formants F1 and F2. Therefore lines 9 and 10 may be quantised with fewer bits than, for example, lines 5 and 6.

By plotting the LSP line frequency locus for a number of TIMIT speech recordings, as shown in Figure 5.11, we can see the line localisation in frequency is fairly limited. The figure shows which lines are located predominantly in frequency regions of less importance to intelligibility: these are natural candidates for being quantised with fewer bits than other lines. The plot was obtained through tenth-order LPC analysis on 40 ms frames with 50% overlap for different male and female speakers. These LPC coefficients were then transformed into LSP values, with the relative frequency of their values computed across 40 analysis bins and then plotted in the vertical axis for each of the LSP lines.

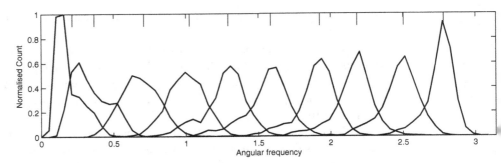

**Figure 5.11** A histogram of the relative frequency of each LSP for tenth-order analysis of speech, showing tick marks at the top of the plot corresponding with the mean angular frequency of the ordered lines.

**Table 5.2.** Average frequency, standard deviation and median frequency for 10 line frequencies.

| No. | Average (Hz) | $\sigma$ (Hz) | Median (Hz) |
|-----|--------------|---------------|-------------|
| 1   | 385          | 117           | 364         |
| 2   | 600          | 184           | 727         |
| 3   | 896          | 241           | 1091        |
| 4   | 1269         | 272           | 1455        |
| 5   | 1618         | 299           | 1818        |
| 6   | 1962         | 306           | 2182        |
| 7   | 2370         | 284           | 2545        |
| 8   | 2732         | 268           | 2909        |
| 9   | 3120         | 240           | 3272        |
| 10  | 3492         | 156           | 3636        |

Table 5.2 lists the average and median LSP frequency for a flat spectral frame. It also lists the standard deviation between the average frequency of each line and its actual location (averaged over several hundred seconds of speech extracted randomly from the TIMIT database and processed with tenth-order analysis). In the table, the standard deviation of lines 1 and 10 is significantly smaller than that of other lines, such as line 6. In a non-uniform scalar quantisation scheme, this would mean that lines 1 and 10 require fewer quantisation bits than line 6 for equivalent signal-to-quantisation noise ratio.

Similarly to Figure 5.11, a differential analysis has been plotted in Figure 5.12, where the relative frequency of frame-to-frame difference is computed for each line. Again, it can be seen that the frame-to-frame variation for individual lines is localised. Both Figure 5.11 and Figure 5.12 corroborate the evidence of Table 5.2 in showing that lines

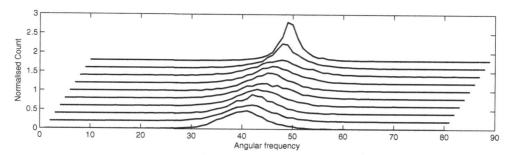

**Figure 5.12** A histogram of the relative frequency difference between LSP line locations for consecutive analysis frames, plotted using 80 analysis bins.

1 and 10 exhibit less variation in their absolute value, but also in their frame-to-frame variation than do the centre lines.

Experimental results suggest that dynamically choosing one or other of the two quantiser bases (frame-to-frame differential, or distance-from mean) to quantise and transmit can improve quantiser performance [9,16]. In practice this implies that the quantised parameters are calculated using both methods for each analysis frame, and the method which results in the closest match is then chosen. A single bit from the quantisation bit pool must then be reserved to flag which method is in use [9].

In a standard CELP coder employing scalar quantisation, around three or four bits are used to quantise each LSP [8,17], or around 30 or 40 bits for a tenth-order representation. A number of techniques have been published in order to allow representation with fewer bits at equivalent levels of quality. A survey of published results from [4,11] and other previously mentioned references can assess the spectral distortion associated with a number of different coding methods at between 3 and 3.6 bits per LSP. This is shown in Figure 5.13 which plots the spectral distortion due to quantising a tenth-order LSP representation using the following methods:

- uniform scalar quantisation (UQ and LSF);
- differential scalar quantisation (DQ and LSFD, also known as DF);
- adaptive forward sequential quantisation (AQFW);
- adaptive backward sequential quantisation (AQBW);
- non-uniform scalar quantisation (NUQ);
- dynamically interpolated quantisation (LSFI);
- dynamically interpolated differential quantisation (LSFID).

From the figure it can be seen that all methods improve at a similar rate as bit allocation increases, and that differential methods are better than non-differential methods. Furthermore non-uniform quantisation improves over uniform quantisation. It should

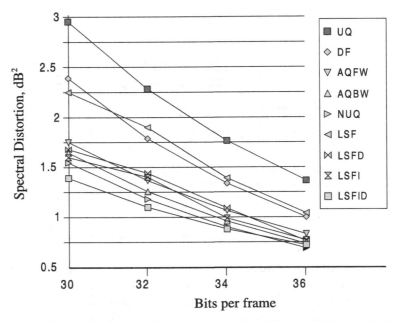

**Figure 5.13** A plot of average spectral distortion due to nine different published methods of LSP quantisation for allocations of between 30 and 36 quantisation bits per frame.

be noted that the results shown derive from some slightly different conditions and so the reader should be wary about making direct comparisons (as opposed to relative comparisons).

Vector quantisation methods can be used to improve over scalar quantisation [8]. In VQ, a candidate vector is generated comprising the set of parameters to be quantised. The vector is compared, in turn, to each entry in a codebook of equal sized vectors. The distance between the candidate vector and each codebook entry is calculated, usually based on Euclidean distance, or sometimes on a perceptually relevant distance measure (which we will explore in Section 7.2). The index of the codebook entry nearest the candidate vector is then taken as the transmitted parameter.

Split VQ allows the vector of all LSPs for a particular frame to be split into subvectors, and compared to subcodebooks. A considerable amount of research has been performed to determine optimal vector and codebook splits. A typical split for a tenth-order system is into three subvectors of three lower, three middle and four upper LSPs. In fact this has been shown to be optimal [13] for a three-subvector system, whilst a 4–6 split is optimal for a two-subvector system.

The reader should also be aware that this has been an extremely hot research area in recent years, and that some quite impressive results have been achieved which exceed even the performance of the methods presented here.

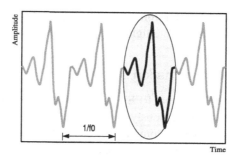

**Figure 5.14** A pitch waveform showing several individual pitch pulses. One is circled, and the spacing between pulses, shown, determines the perceived pitch frequency.

## 5.3    Pitch models

The source-filter model is perhaps the ultimate in speech parameterisation, with different processing blocks dedicated to replicating the effects of the human vocal system: LPC/LSP for the vocal tract, random noise (and similar) for the lung excitation, and a pitch filter or similar to recreate the effect of the glottis.

Measurements of the human pitch-production system, especially those using microwave and X-ray sensors, reveal the action of the glottis, which is not a smooth action: it does not generate a pure sinewave tone. In actual fact, the pitch waveform is made up of a sequence of very spiky pulses. This is shown in Figure 5.14 where one pulse has been identified from a sequence of several plotted as if isolated from a speech utterance.

There has been quite a lot of research on determining pitch shapes: how these relate to overall vocal quality, speech intelligibility, and so on. There is substantial evidence that the fidelity of the pitch pulse shape is important to overall perceived quality, and other evidence to indicate that the specific pulse shapes, which vary considerably from person to person, are one of the differentiating factors for speaker recognition (where an automatic system identifies someone through their voice, see Section 7.5).

When coding or compressing speech in a parametric fashion, there are several items of information that are important for pitch, and these are handled differently by the various speech compression algorithms. These are listed below:

- the actual shape of the pulse;
- the relative heights/locations of the negative- and positive-going spikes;
- the amplitude of the largest spike;
- the spacing between pulses.

The highest quality compression algorithms would consider all aspects. Some code only the bottom three items, CELP coders tend to code the bottom two, and regular-pulse excitation systems code only the bottom one. It goes without saying that more bits are required to code more information, thus many algorithms give priority to only the most important aspects for intelligibility, namely the lower entries on the list.

## 5.3.1    Regular pulse excitation

Regular Pulse Excitation (RPE) is a parametric coder that represents the pitch component of speech. It is most famously implemented in ETSI standard 06.10, and currently is the primary mobile speech communications method for over a third of the world's population, by any measure an impressive user base. This is due to its use in the GSM standard, developed in the 1980s as a pan-European digital voice standard. It was endorsed by the European Union, and quickly found adoption across Europe and then beyond.

GSM codes frames of 160 13-bit speech samples (at a sampling rate of 8 kHz) into 260 compressed bits. A decoder takes these and regenerates 160-sample output speech frames. There are many sources of information on GSM, not least the open standard documents, so there is no need to consider full details here. However we will examine the pitch coding system for GSM 06.10, the traditional or 'full rate' standard.

In GSM, the original speech is analysed to determine vocal tract parameters (LPC coefficients) which are then used to filter the same vector of 160 speech samples to remove the vocal tract information, leaving a residual. The eight LPC coefficients will be transformed into LARs (Log Area Ratios) for transmission.

The residual is then split into four subframes. Each subframe is analysed separately to determine pitch parameters. The analysis is made on the current subframe concatenated with the three previous reconstituted subframes. The reconstituted subframes are those that have been generated from the previous pitch values – those that have been quantised for transmission. Thus they are effectively the subframes as generated by a decoder.

These four subframes (the current one, and the three reconstituted ones) form a complete frame which is subjected to long-term prediction (LTP) which is actually quite simple, and will be discussed in the next section. When this contribution is removed from each subframe a set of pitch-like spikes remain – assuming of course the presence of pitch in the original speech. An RPE analysis engine compares the subvector of spikes to four candidates, one of which is chosen, along with a location (grid position) to represent the pitch spikes in that subframe.

This pulse train is actually coded by ADPCM before transmission. This entire coding process is known as RPE-LTP, and is shown diagrammatically in Figure 5.15. If there were no pitch in the original speech (it has been judged to be unvoiced speech), then the residual is represented as random noise instead.

Up to 13 pitch pulses are coded per 40-sample subframe, achieved through downsampling at a ratio of 1:3 from several sequence start positions 1, 2 or 3. As can be imagined a set of regular pulses is not particularly similar to the pitch waveform shown in

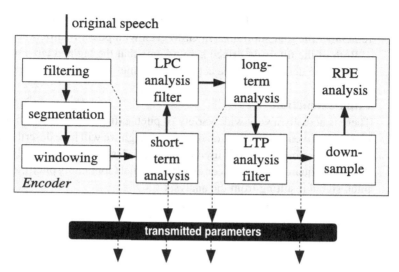

**Figure 5.15** GSM RPE-LTP encoder block diagram, showing transmitted parameters.

Figure 5.14, but is perfectly sufficient, allied with the LTP and LPC analysis, to maintain speech intelligibility.

## 5.3.2    LTP pitch extraction

We can define a long-term prediction pitch filter by a number of taps and lags. We will illustrate for a simple one-tap pitch filter, which is the most common variant. Given a vector $c$ of audio samples, we can add in a pitch component identified by an amplitude $\beta$ and lag $M$ as shown in Equation (5.33) to generate a spiky signal $x$ which includes pitch:

$$x(n) = c(n) + \beta x(n - M).\tag{5.33}$$

$\beta$ scales the amplitude of the pitch component and the lag $M$ corresponds to the primary pitch period. For more complex pitch representations, more taps (with different lags) are used. It is also possible for multiple taps to be fixed around a single lag value to better approximate the shape of the pitch pulse. A three-tap filter with single lag is shown in Equation (5.34):

$$x(n) = c(n) + \beta_1 x(n - M - 1) + \beta_2 x(n - M) + \beta_3 x(n - M + 1).\tag{5.34}$$

Often, fractional pitch lags are used to improve pitch quality further. In all cases, the LTP filter, being IIR, calculates using past output values. $M$ can range in length from less than a subframe to more than a frame, so at times the LTP filter evidently acts on a long sequence of audio (hence the name 'long-term'). This may be problematic in implementation since the pitch can change faster than the other vocal characteristics: remember the four subframes in GSM each with possibly different pitch values.

Both of the filters described here assume that the lag and tap amplitudes are already known. The derivation of the actual pitch values is named pitch extraction.

### 5.3.2.1    Pitch extraction

There is actually a very wide variety of pitch extraction methods in published literature (some are given later in Section 6.2.1), although we will here describe one of the simpler, and more common methods used in speech coders. This method relies upon minimising the mean-squared error between an LPC residual (containing pitch), and the reconstructed pitch signal resulting from the analysis.

If $E$ is the mean-squared error, $e$ is the residual and $e'$ is the reconstructed pitch signal after analysis, then:

$$E(M,\beta) = \sum_{n=0}^{N-1} \{e(n) - e'(n)\}^2 \tag{5.35}$$

and assuming a single tap pitch filter as in Equation (5.33) then:

$$E(M,\beta) = \sum_{n=0}^{N-1} \{e(n) - \beta e(n-M)\}^2 \tag{5.36}$$

where $N$ is the analysis window size (usually one or more subframes). In order to find the set of $\beta$ and $M$ that minimises the mean-squared error (i.e. best reproduces the original pitch signal) then we need to differentiate the expression and set to zero:

$$\frac{\delta E}{\delta \beta} = \sum_{n=0}^{N-1} \{2\beta e^2(n-M) - 2e(n)e(n-M)\} = 0 \tag{5.37}$$

so

$$\beta_{\text{optimum}} = \frac{\sum_{n=0}^{N-1} e(n)e(n-M)}{\sum_{n=0}^{N-1} e^2(n-M)}. \tag{5.38}$$

We can now substitute the optimum $\beta$ from Equation (5.38) into (5.36) to give the optimum $M$ from:

$$E_{\text{optimum}}(M) = \sum_{n=0}^{N-1} e^2(n) - E'_{\text{optimum}}(M) \tag{5.39}$$

as only the second part of the equation varies with respect to $M$ it must be maximised to minimise the error. Thus the following must be determined with respect to each permissible value of $M$, and the value at which a maximum occurs, stored:

$$E'_{\text{optimum}}(M) = \frac{[\sum_{n=0}^{N-1} e(n)e(n-M)]^2}{\sum_{n=0}^{N-1} e^2(n-M)}. \tag{5.40}$$

Effectively, this means that the best pitch delay (and the one that is finally chosen) is the one that, averaged over a whole analysis frame, allows the best prediction of that subframe. Once found, the pitch scaling factor $\beta$ is then chosen similarly, as the optimal scaling factor averaged over the analysis frame:

$$\beta = \frac{\sum_{n=0}^{N-1} e(n)e(n-M)}{\sum_{n=0}^{N-1} e^2(n-M)}. \tag{5.41}$$

This can be illustrated by the following MATLAB function which performs these same equations in a fairly inefficient, but straightforward fashion:

```
function [B,M]=ltp(sp)
    n=length(sp);
    %Establish upper and lower pitch search limits
    pmin=50;    pmax=200;
    sp2=sp.^2; %pre-calculate to save time
    for M=pmin:pmax
        e_del=sp(1:n-M);
        e=sp(M+1:n);
        e2=sp2(M+1:n);
        E(1+M-pmin)=sum((e_del.*e).^2)/sum(e2);
    end
    %Find M, the optimum pitch period
    [null, M]=max(E);
    M=M+pmin;
    %Find B, the pitch gain
    e_del=sp(1:n-M);
    e=sp(M+1:n);
    e2=sp2(M+1:n);
    B=sum(e_del.*e)/sum(e2);
```

It can be very interesting to plot the E that is calculated just before the maximum is found. Normally, the plot will show a periodic structure (and illustrates the issue of pitch doubling or halving, if the pitch search range is reduced to exclude the largest peak then very often the second largest peak will be at twice the pitch period, as discussed further in Section 5.3.3).

For computational purposes, note the identical denominator in Equations (5.40) and (5.41) which would only have to be determined once in an efficient implementation. Although the MATLAB code attempts a slight saving in computation by pre-calculating an array of squared input samples, it would actually be more efficient if we performed the correlation by shifting the numerator $e(n)$ only, rather than both the numerator and the denominator. In fact this method is used in most real-time speech coders. It is left as an exercise for the reader to modify the MATLAB function in that way and see whether the same results can be found.

### 5.3.3    Pitch issues

The pitch extraction method of Section 5.3.2.1, in common with many other methods, often produces an answer equal to half, or twice of the actual pitch period. This is called *pitch halving* and *pitch doubling* and is the scourge of many engineers working with pitch detection algorithms.

To some extent, setting hard limits on pitch period can provide an answer (i.e. saying, for example, that pitch cannot be less than 50 Hz or more than 300 Hz), but such a range still has ample scope for doubling or halving. Just think of the comparison between a squeaky five year old child's voice and that of a deep bass voice such as those belonging to Paul Robeson or Luciano Pavarotti. Many algorithms do not impose absolute limits, but disallow sudden shifts in pitch as being unlikely in a real scenario. True of speech, but such decisions tend to result in speech systems unable to handle music, singing or DTMF (Dual Tone Multiple Frequency) and facsimile signalling.

G.728 for example, limits the pitch slew rate between subframes to ±6 samples unless the relative strength of the new pitch component is at least 2.5 times greater than that of the previous frame [18]. Any limits used are likely to require empirical testing with a range of subject matter.

## 5.4    Analysis-by-synthesis

The idea behind analysis-by-synthesis at the encoder is to analyse a frame (or more) of speech, and extract parameters from this. These parameters are then used to create a frame of reconstructed speech. The frames of original and reconstructed speech are then compared to see how closely they match. Some part of the parameter extraction process is then varied to create a slightly different set of parameters, which are in turn compared to the original speech.

Perhaps several hundred iterations are made across a search space, and the best set of parameters (based on how close the match is between original and reconstructed speech) are then transmitted to the receiver. Something to consider is that the parameters may need to be quantised before being transmitted to the decoder. In this case the quantised-dequantised parameters are the ones used by the encoder to check how good is the matching.

Before we look at the most famous of the analysis-by-synthesis coding structures, it is important to remember that 'degree of matching', calculated as a difference between vectors, may not relate at all to how a human perceives degree of difference. As a very trivial example, imagine a continuous sinewave original signal. Next imagine a version which is delayed by a few degrees. The degree of matching in a mean-squared sense will be quite low – in fact it may be less than the matching between the original sinewave and random noise. However the perceived difference between the sinewaves is probably zero, but huge when one signal is random noise.

Therefore, most practical analysis-by-synthesis algorithms use a perceptual matching criterion: either a perceptual weighting filter or something like a spectral distortion measure (see Section 3.3.2), rather than a mean-squared match.

## 5.4.1 Basic CELP

CELP is the logical culmination of an evolutionary process in speech compression algorithms: it can provide excellent quality speech at low bitrates and is a common choice for speech products. It utilises a source filter model of speech, parameterised as we have seen with gain, vocal tract, pitch and lung excitation information.

CELP stands for either *Code Excited Linear Prediction* or *Codebook Excited Linear Prediction* depending on whom you ask. What is certain, though, is that the technique collectively describes quite a variety of similarly structured algorithms. We will begin with the basic structure, and subsequently look briefly at algebraic, adaptive and split variants. Be aware that this has been an area of intense research activity for over a decade now: many mutant forms have emerged in the research literature.

We will start with the basic CELP encoder, designed to decompose a speech signal into various parameters. A block diagram of such a system is shown in Figure 5.16. This shows the speech signal being filtered (including normalisation, yielding gain information), segmented, and windowed. It is then analysed for pitch components (represented by LTP parameters) and vocal tract resonances (represented by LPC coefficients). Readers may note the similarity to the RPE system in Figure 5.15, and indeed both coders do share many characteristics.

Where the CELP and RPE systems differ most greatly is in the handling of the original lung excitation signal. RPE treats this as either white Gaussian noise, or as a pulse-train. The CELP coder takes a different approach: it utilises a large codebook of *candidate vectors* at both encoder and decoder, and essentially runs through an iterative process to attempt to identify which of the candidate excitation vectors best represents the actual lung excitation.

At least that is the theory – in practice none of the parameters exactly characterise the required information perfectly. This means that both the LPC and LTP representations, neither being perfect, will 'leak' information. In the RPE encoder, vocal tract information not caught by the LPC analysis is unlikely to be picked up by the LTP and RPE analysis, and so that information will be lost to the encoder, and consequently not transmitted to the decoder. This contributes to loss of quality in speech processed by such a system.

In CELP, the codebook of candidate excitation vectors can often pick up some of the information which was not caught by the LTP and LPC analysis. So in practice the codebook does not just model lung excitation. Since this mechanism greatly improves

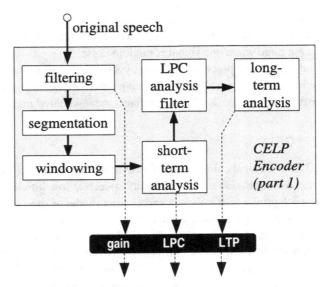

**Figure 5.16** A block diagram of part of a CELP encoder, showing original speech being decomposed into gain, LPC and LTP parameters.

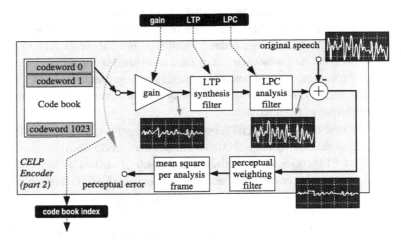

**Figure 5.17** A block diagram of the remainder of the CELP encoder. Gain, LPC and LTP parameters were obtained in the first part shown in Figure 5.16, whilst the section now shown is devoted to determining the optimum codebook index that best matches the analysed speech.

the quality of the system over RPE, we will consider in a little more detail exactly how it works.

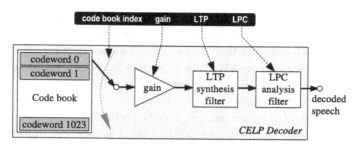

**Figure 5.18** A block diagram of the remainder of the CELP decoder utilising codebook index, gain, LPC and LTP parameters to recreate a frame of speech.

Following a particular analysis frame of speech through the CELP encoder, first the basic gain, pitch and vocal tract parameters are determined (shown in Figure 5.16), and then these parameters are used to recreate pseudo-speech as in Figure 5.17, as the output from the LPC analysis filter. The first candidate vector in the codebook, named codeword 0, is used as the lung excitation. Amplification, LPC and LTP synthesis filters add gain, pitch and vocal tract information to the lung excitation in order to derive a frame of pseudo-speech.

This pseudo-speech is compared to the original frame of speech. In fact the comparison simply finds a difference vector between the two, perceptually weights this (something we will return to later in Section 7.2), and calculates the mean square for that frame: a single perceptual error value for the current input speech frame. The process is now repeated for codeword 1, and again results in a single perceptual error value. Each element in the codebook is now tried in turn.

For the typical codebook shown, the result will be 1024 perceptual error values. Each one is a measure of the difference between the pseudo-speech recreated with that codebook index and the original speech. Thus the codebook index which resulted in the smallest perceptual error value is the one which can be used to best represent the original speech, and this index (0 to 1023) is transmitted from encoder to decoder.

At the decoder, shown in Figure 5.18, the transmitted parameters are used to recreate a frame of decoded speech. The codeword selected in the decoder is that identified by the transmitted codebook index. This codeword is identical to that at the same position in the encoder – and is thus guaranteed to be the best of the candidate vectors. We would therefore expect the frame of decoded speech to be similar to the original analysed speech. It is evident that the decoder is considerably simpler than the encoder – at least 1024 times in the example shown (since it does not need to repeat once for each codebook entry, but just for the indicated codebook entry). This is offset somewhat by the trend to apply post-filtering to the decoder output (not shown here) in order to improve audio quality.

It is clear that since the codebook is the differentiating factor in the CELP method, we will explore that in a little more detail. First however, we need to remind ourselves of the need for quantisation: the raw LPC, LTP and gain parameters need to be quantised in some way. As we have seen in Section 5.2.5, LPC parameters are rarely used as-is. Within CELP coders they are generally transformed to line spectral pairs prior to being output from the encoder.

In fact all of the parameters, with the probable exception of the codebook index, will be quantised, and in the process transformed in some way. Remembering that the encoder incorporates the decoding process in its codebook search loop, it is important to note that the actual parameters used in this part of the encoder are *already quantised* and then dequantised. The main reason is that if the encoder uses unquantised parameters it may well find a different candidate excitation vector to the one it would choose if operating on quantised-dequantised parameters. Since the actual speech output from the decoder has access only to the quantised-dequantised parameters, then the encoder must use the same values to ensure the best possible speech is generated.

### 5.4.1.1    CELP codebooks

As mentioned previously, each codebook is populated by a number of codewords. These are used as candidate vectors within the encoder, where each candidate is examined in turn, and the candidate that results in the best matching speech frame is chosen. For a typical system that analyses speech in 20 ms frames, and has a sample rate of 8 kHz, the candidate vectors need to consist of $8000 \times 0.02 = 160$ samples. Quite often, pitch is analysed and represented in subframes that may be 5 ms long – four subframes per frame – and thus the LTP parameters change four times as often as the LPC parameters, but otherwise the processing structure remains unchanged.

In the original CELP technique [19,20], the candidate vectors in each codebook were generated from a random number generation algorithm – 'seeded' identically at encoder and decoder so that they would contain exactly the same vectors: at the most basic level, a 1024 set of $1 \times 160$ random numbers. More modern variants will introduce some structure into the codebook [12] – or allow the addition of two vectors to be used as an excitation candidate. Such techniques are known as split codebooks.

Many useful enhancements to basic CELP rely on the fact that the CELP encoder (Figure 5.17) actually contains a decoder as part of its structure. That means that, for the same codebook index, the encoder pseudo-speech is identical to the decoder output speech. This is exploited in allowing the encoder to predict the state of the decoder, something necessary for any adaptive system. One such system allows the codebook to adapt: ensuring that both encoder and decoder codebooks adapt equally (keep in step) is tricky, but necessary to ensure performance.

Although speech quality improvement has always been the main driving factor behind the advance of the CELP technique, computational complexity reduction has been another significant factor. A third factor has been minimisation of processing latency. Three well-known enhancements to CELP are now discussed that address the quality, computational complexity, and processing latency issues.

## 5.4.2    Algebraic CELP

As mentioned previously, the computational complexity of the CELP coder is large: this is due to the need to synthesise pseudo-speech for every vector within the codebook. This already-complicated process must be repeated hundreds or thousands of times for each analysis frame. Whilst many methods attempt to reduce the amount of computation in the CELP coder by reducing the number of loops (see Section 5.4.3), algebraic CELP (often known simply as ACELP) attempts to reduce the amount of computation required within each loop.

Referring to Figure 5.17, the action required for each excitation vector is to amplify it, perform LTP synthesis (to put in the pitch component) and perform LPC synthesis (to add in the vocal tract components). Each of these are linear time-invariant (LTI) processes which means that mathematically it does not matter in which order they are performed. However in terms of processing complexity, it is often better to perform the amplification last (so that the numbers being processed by the filters do not become too large to be easily represented), and we will demonstrate that, in ACELP at least, it is better to perform the LPC filter first.

The filtering process, as described in Section 2.3, consists of a series of multiplications. Filter coefficients are multiplied by input values, and summed to produce output values. In the CELP encoder, this is done twice for each codeword – first in the LPC filter and secondly in the LTP filter. These may have 10 and two coefficients respectively, so clearly the LPC filter is the most computationally complex one. If it were possible to reduce the complexity of the LPC filter, there would be a corresponding large reduction in overall complexity.

In ACELP, the codewords are carefully constructed so that they are sparse (meaning most elements are zero, with a few individual non-zero elements scattered around), with each element in the vector being constrained to $+1$, $0$ or $-1$. These codewords are then fed directly into the LPC synthesis filter. Let us remind ourselves what the filter looks like mathematically:

$$y(n) = a(1) \times x(n) + a(2) \times x(n-1) + a(3) \times x(n-2) + \cdots + a(m+1)$$
$$\times x(n-m).$$

In this case, $x()$ contains the input data, $y()$ contains the output data, and the filter coefficients are $a()$. If we rewrite the filter on the basis of some illustrative input vector $x = [+1, 0, 0, -1, 0, 0, -1]$, meaning $x(1) = +1$, $x(2) = 0$, $x(7) = -1$ and so on, then the filter equation reduces to:

$$y(n) = a(1) + 0 + 0 - a(4) + 0 + 0 - a(7).$$

This equation involves no multiplications, simply a summation of all non-zero elements (and being sparse there are few numbers that need to be summed). ACELP generally operates with 80% of the elements in the codebook set to zero. This is significantly simpler, and faster to calculate than the full filter equation that would be required with a codebook vector of random elements.

A further advantage of ACELP is that it is entirely possible to calculate the position (and value) of each non-zero element in the candidate vector on-the-fly. Such an action requires some minor additional processing to generate the vectors, but does mean that the static codebook storage is no longer required. For memory-constrained embedded systems, this may well be a significant consideration.

### 5.4.3     Split codebook schemes

Unlike ACELP (above, Section 5.4.2) which reduces the computational complexity of the LPC filtering process, split codebook schemes attempt to reduce the 'search space'. By default the search space is the entire codebook since each element in the codebook is tested in turn, and there may be very many such elements. If there were some way to reduce the number of candidate excitation vectors tested, then complexity can be reduced.

Two main methods exist. The first one is to order the codebook in such a way that the codewords are arranged by known approximate characteristic. Through one of several different predictive processes, it is possible to determine the rough spectral or time-domain characteristic of the vector required for a particular speech analysis frame. Only codewords fitting the required description need be tested as candidates. Occasionally such a system may make a suboptimal choice, but given a sufficiently good predictive algorithm, such events may be rare.

The second method is to use two or more different codebooks with orthogonal properties. Each orthogonal property is found separately, and the two codebooks can be searched independently. As an example of a dual-codebook system, codebook 1 – perhaps carrying a selection of spectral distributions – is searched with codebook 2 set to some arbitrary value. Once the best match from codebook 1 is found, this is maintained whilst codebook 2 – perhaps carrying temporal envelope information – is searched. Ideally the two are orthogonal (meaning that whichever codeword from one codebook is used, the best match from the other codebook will not change for a particular frame of analysed speech).

At one extreme, a split system could be a mixed excitation linear prediction (MELP) coder – although this name is typically reserved for coders with split excitation that do not utilise a codebook (and hence do not perform analysis-by-synthesis) [21]. VSELP (vector sum excited linear prediction [22]) is a common split system with ordered codebooks that are searched independently, with the best codeword from each added together to provide the excitation vector. In this case, each codebook output is subject to an individual scaling factor. Less important for this discussion, but a fact which should be mentioned for consistency, is that in the VSELP coder one of those codebooks is used to replace the LTP pitch filter. This codebook is pitch-adaptive, and is obviously searched, and optimised, before the other two codebooks of candidate vectors are searched.

### 5.4.4     Forward–backward CELP

When using a speech coder such as CELP in a real application, it will naturally take some time for a particular sound of speech to travel from the input microphone, through the system, and arrive at the output loudspeaker. Apart from the transmission of speech through wiring (or fibre optics, or wireless), processing takes some finite time.

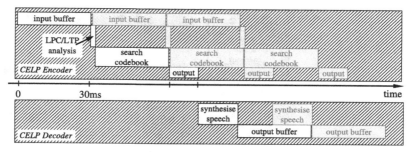

**Figure 5.19** Timing diagram for basic CELP encoder and decoder processing, illustrating processing latency.

The first delay comes about due to collection of audio samples into a buffer before they are processed: for a speech system working on a 30 ms analysis frame at a sample rate of 8 kHz, each frame contains $8000 \times 0.03 = 240$ samples. Processing can typically not begin until all of those 240 samples are available, and naturally the final sample in the buffer was collected 30 ms after the first sample. Even with no other processing, such a system will delay audio by 30 ms. The output buffering arrangements might well affect the system in a similar way. These latencies between input and output ignore the operation of any coder, decoder, and all propagation delays.

Next we look at the operation of the CELP encoder of Figure 5.17. Looping around 1024 times, an entire decoding operation must be performed, and then perceptual weighting and mean-squared calculation. None of this can begin until the sample input buffer has been filled, and then the following process is likely to require a significant amount of processing time. This does, of course, depend on the clock and calculation speed of the underlying hardware. To put this into perspective, the original inventors of CELP found that their Cray-1 supercomputer required 125 seconds of processing time to process just a single second of speech [20]. Of course, computers are far more powerful today than in the 1980s; however such processing is still very far from instantaneous.

As CELP began to be adopted in real systems, figures of 200–300 ms latency were observed. Unfortunately at these latencies, human conversations become rather strained: people begin to speak over one another, and feel uncomfortable with the long pauses in conversation. A useful figure of merit is that most people will not notice a latency of 100–150 ms, but beyond this it starts to become intrusive to conversation.

Clearly, a reduced latency CELP was required: both for the input/output buffering, and for the processing times. The solution was found, and standardised as ITU G.728, boasting submillisecond processing latency. The primary technique used to achieve the latency reduction was the forward–backward structure.

Despite the unusual name, forward–backward CELP refers to the order in which processing is performed. This is perhaps best illustrated in a timing diagram. First we

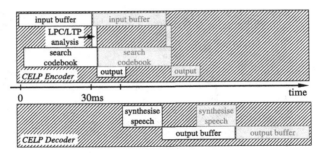

**Figure 5.20** Timing diagram for forward–backward CELP encoder and decoder processing.

consider the basic CELP timing structure as shown in Figure 5.19. It can clearly be seen that speech has to be gathered (buffered), then analysed for gain, LPC and LTP parameters, before these are used in the lengthy codebook search loop. Once this is complete, the previously determined parameters, and codebook index can be output for transmission. The decoder receives these, uses them to synthesise speech, which it then outputs. Quite clearly, the latency from input to output is dominated by the codebook search loop.

By contrast the forward–backward CELP structure, illustrated in Figure 5.20 manages to perform the codebook search in parallel with the speech buffering. This is an interesting concept, only possible because the encoder uses the gain, LPC and LTP parameters found in the *previous* speech frame for the synthesis part of the encoder. The diagram clearly illustrates the significant reduction in processing latency that this provides.

Note that the actual amount of processing is unchanged – it is just organised differently so that the output speech can be made available earlier. In practice, if latency was a particular issue, the frame size would be reduced to 20 ms and a system of overlapping frames used (with half of each analysis frame already having been collected earlier, the buffering delay is reduced).

## 5.5    Summary

This chapter presented a survey of the main techniques in speech communications – primarily related to the quantisation of the speech signal, and the trade-offs that this involves in terms of quality and computational complexity. Both time-domain waveform quantisation and parameterisation were discussed, before a presentation of pitch modelling and extraction techniques. Finally, the important and pragmatic audio engineering subject of analysis-by-synthesis was explored.

Next, the major remaining methods of audio analysis (that have not already been covered) will be presented in Chapter 6, before Chapter 7 rounds off with advanced topics that make use of many of the techniques discussed in the current and previous chapters.

---

**Infobox 5.3** Speech coding standards based around CELP and RPE

---

There is an entire spectrum of speech coding and compression systems, and then another entire spectrum designed to handle music and general audio. The tendency is for the higher rate algorithms to have higher perceived quality – but there are also significant differentiating factors in the computational complexity, and in the underlying techniques used: the class of CELP-based systems contains many weird and wonderful performance-enhancing ideas in addition to the basic codebook excited linear prediction structure.

| Name | Description |
|---|---|
| G.723.1 | The basic ITU CELP standard at 5.3 or 6.3 kbits/s |
| G.728 | Low delay CELP, 16 kbits/s sampled at 8 kHz |
| G.729 | Algebraic CELP 8 kbits/s standard |
| TETRA | CELP 4.8 kbits/s |
| FS1015 | LPC-based tenth-order system at 2.4 kbits/s |
| FS1016 | Basic CELP at 4.8 or 7.2 kbits/s |
| GSM 06.10 | Regular pulse excited, 13.6 kbits/s |
| MELP | Mixed excitation CELP 1.6 to 2.4 kbits/s |
| QCELP | Basic CELP at 8 and 13 kbits/s |

Many variants and extensions exist to these original standards.

## Bibliography

- *Principles of Computer Speech*
  I. H. Witten (Academic Press, 1982)

- *Linear Prediction of Speech*
  J. Markel and A. Gray (Springer-Verlag, 1976)

- *Advances in Speech, Hearing and Language Processing*
  Ed. W. A. Ainsworth (JAI Press, 1990)

- *Spoken Language Processing: A Guide to Theory, Algorithm and System Development*
  X. Huang (Prentice-Hall, 2001)

- *Speech and Audio Signal Processing: Processing and Perception of Speech and Music*
  B. Gold and N. Morgan (Wiley, 1999)

- *Fundamentals of Speech Signal Processing*
  S. Saito and K. Nakata (Academic Press, 1985)

- *Digital Processing of Speech Signals*
  L. R. Rabiner and R. W. Schafer (Prentice-Hall, 1978)

- *Computer Speech Processing*
  F. Fallside and W. Woods (Prentice-Hall, 1985)

- *The Acoustics of Speech Communication: Fundamentals, Speech Perception Theory, and Technology*
  J. M. Pickett (Allyn and Bacon, 1998)

# References

[1]  J. Markel and A. Gray. *Linear Prediction of Speech*. Springer-Verlag, 1976.

[2]  J. Makhoul. Linear prediction: A tutorial review. *Proc. IEEE*, **63**(4): 561–580, 1975.

[3]  S. Saito and K. Nakata. *Fundamentals of Speech Signal Processing*. Academic Press, 1985.

[4]  N. Sugamura and N. Favardin. Quantizer design in LSP speech analysis-synthesis. *IEEE J. Selec. Areas Comms*, **6**(2): 432–440, 1988.

[5]  S. Saoudi, J. Boucher, and A. Guyader. A new efficient algorithm to compute the LSP parameters for speech coding. *Signal Proc.*, **28**(2): 201–212, 1995.

[6]  TI and MIT TIMIT database. *A CD-ROM database of phonetically classified recordings of sentences spoken by a number of different male and female speakers*, disc 1–1.1, 1990.

[7]  N. Sugamura and F. Itakura. Speech analysis and synthesis methods developed at ECL in NTT – from LPC to LSP. *Speech Commun.*, **5**: 213–229, 1986.

[8]  J. S. Collura and T. E. Tremain. Vector quantizer design for the coding of LSF parameters. In *Proc. Int. Conference on Acoustics, Speech and Signal Processing*, pages 29–32, 1993.

[9]  I. V. McLoughlin. LSP parameter interpretation for speech classification. In *Proceedings of the 2nd IEEE International Conference on Information, Communications and Signal Processing*, December 1999.

[10]  I. V. McLoughlin and F. Hui. Adaptive bit allocation for LSP parameter quantization. In *Proceedings of the IEEE Asia-Pacific Conference Circuits and Systems*, number 231, December 2000.

[11]  Q. Zhao and J. Suzuki. Efficient quantization of LSF by utilising dynamic interpolation. In *IEEE International Symposium on Circuits and Systems*, pages 2629–2632, June 1997.

[12]  European Telecommunications Standards Institute. Trans-European trunked radio system (TETRA) standard. *A European Telecommunications Standards Institute (ETSI) standard*, 1994.

[13]  K. K. Paliwal and B. S. Atal. Efficient vector quantization of LPC parameters at 24 bits per frame. In *Proc. Int. Conf. on Acoustics, Speech and Signal Processing*, pages 661–664, 1991.

[14]  D. Chang, S. Ann, and C. W. Lee. A classified vector quantization of LSF parameters. *Signal Proc.*, **59**(3): 267–273, June 1997.

[15]  R. Laroia, N. Phamdo, and N. Farvardin. Robust and efficient quantization of speech LSP parameters using structured vector quantizers. In *Proc. Int. Conf. on Acoustics, Speech and Signal Processing*, pages 641–644, 1991.

[16]  H. Zarrinkoub and P. Mermelstein. Switched prediction and quantization of LSP frequencies. In *Proc. Int. Conf. on Acoustics, Speech and Signal Processing*, pages 757–760, 1996.

[17]  C. S. Xydeas and K. K. M. So. A long history quantization approach to scalar and vector quantization of LSP coefficients. In *Proc. Int. Conf. on Acoustics, Speech and Signal Processing*, pages 1–4, 1993.

[18]  J.-H. Chen, R. V. Cox, Y.-C. Lin, N. Jayant, and M. J. Melchner. A low-delay CELP coder for the CCITT 16 kb/s speech coding standard. *IEEE J. Selec. Areas Comms*, **10**(5): 830–849, 1992.

[19]  B. S. Atal. Predictive coding of speech at low bitrates. *IEEE Trans. Commun.*, COM30: 600–614, 1982.

[20]  M. R. Schroeder and B.S. Atal. Code-excited linear prediction CELP: High-quality speech at very low bit rates. In *Proceedings of the International Conference on Acoustics, Speech and Signal Processing*, pages 937–940, 1985.

[21]  L. M. Supplee, R. P. Cohn, J. S. Collura, and A. V. McCree. MELP: The new Federal standard at 2400 bps. In *IEEE International Conference on Acoustics, Speech and Signal Proc.*, volume 2, pages 1591–1594, April 1997.

[22]  I. A. Gerson and M. A. Jasiuk. Vector sum excited linear prediction (VSELP) speech coding at 8 kbps. In *IEEE International Conference on Acoustics, Speech and Signal Proc.*, volume 1, pages 461–464, April 1990.

# 6    Audio analysis

Analysis techniques are those used to examine, understand and interpret the content of recorded sound signals. Sometimes these lead to visualisation methods, whilst at other times they may be used in specifying some form of further processing or measurement of the audio.

There is a general set of analysis techniques which are common to all audio signals, and indeed to many forms of data, particularly the traditional methods used for signal processing. We have already met and used the basic technique of decomposing sound into multiple sinusoidal components with the Fast Fourier Transform (FFT), and have considered forming a polynomial equation to replicate audio waveform characteristics through linear prediction (LPC), but there are many other useful techniques we have not yet considered.

Most analysis techniques operate on analysis windows, or frames, of input audio. Most also require that the analysis window is a representative stationary selection of the signal (stationary in that the signal statistics and frequency distribution do not change appreciably during the time duration of the window – otherwise results may be inaccurate). We had discussed the stationarity issue in Section 2.5.1, and should note that the choice of analysis window size, as well as the choice of analysis methods used, depends strongly upon the identity of the signal being analysed. Speech, noise and music all have different characteristics, and while many of the same methods can be used in their analysis, knowledge of their characteristics leads to different analysis periods, and different parameter ranges of the analysis result.

Undoubtedly, those needing to perform an analysis will require some experimentation to determine the best methods to be used, the correct parameters to be interpreted, and optimal analysis timings.

We will now introduce several other methods of analysing sound that form part of the audio engineer's standard toolkit, and which can be applied in many situations. Following this, we will consider the special cases of pitch analysis for speech, and the subject of understanding how the human brain interprets sound, in auditory scene analysis. Finally we will touch upon the analysis of some other signals such as music and animal noises before we discuss the use of tracking sound statistics as a method of analysis.

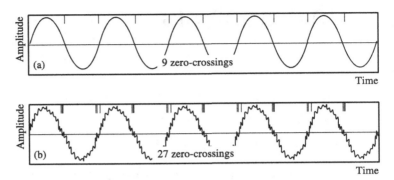

**Figure 6.1** Zero-crossing rate calculation illustrated for a pure sinewave (a) and one corrupted with noise (b). Zero crossings are indicated by tick marks at the top of each plot. The noise-corrupted sinewave exhibits many more zero crossings due to additive noise taking the signal over the trigger level several times each period.

## 6.1     Analysis toolkit

Each of the following methods operates over an analysis frame of size $N$. In many cases, the measures are conducted in the time domain, with the consequence that the frame does not need to be windowed prior to analysis, or for the analysis frames to be overlapped.

### 6.1.1     Zero-crossing rate

There is nothing particularly advanced or complicated about the zero-crossing rate (ZCR). It is a 'poor man's' pitch determination algorithm, working well in the absence of noise, and is designed to be very simple in terms of computation. It works well for a noise-free and simple waveform like a sinewave, the idea being to count how many times the waveform crosses the zero-axis in a certain time: the number of crossings per second will equal twice the frequency. If we define sign{ } to be a function returning $+1$ or 0 depending upon whether the signal is greater than zero or not, then the ZCR for the $i$th analysis frame, of length $N$ can be determined as:

$$\text{ZCR}_i = \frac{1}{N} \sum_{n=0}^{N-1} |\text{sign}\{x_i(n)\} - \text{sign}\{x_i(n-1)\}| . \tag{6.1}$$

The $1/N$ provides the answer as a crossing rate: the proportion of the samples which cross the zero-axis. Most of the time, in a system with fixed frame size, where ZCR values are compared from frame to frame, the division is an unnecessary fixed scaling and therefore not performed.

This is illustrated in Figure 6.1(a) where the zero crossings of a sinewave are counted over a certain analysis time. In this case the fundamental frequency of the sinewave causes nine crossings across the plot. However in the presence of additive noise, the 'wobble' in the signal as it crosses the zero-axis causes several false counts. In Figure 6.1(b) this leads to an erroneous estimate of signal fundamental frequency – in fact an estimate that would be three times too high.

In MATLAB, determining the ZCR is relatively easy, although not particularly elegant:

```
function [zcr]=zcr(segment)
zc=0;
for m=1:length(segment)-1
    if segment(m)*segment(m+1) > 0
        zc=zc+0;
    else
        zc=zc+1;
    end
    zcr=zc/length(segment);
end
```

To illustrate the MATLAB zcr() function above was applied to a recording of speech. The speech was segmented into non-overlapping analysis windows of size 128 samples, and the ZCR determined for each window. The results, plotted in Figure 6.2, show a good correspondence between the ZCR measure and the frequencies present in the speech – higher frequency regions of the recorded speech, such as the /ch/ sound, have a higher ZCR measure.

A pragmatic solution to the problem of noise is to apply a threshold about the zero-axis. In essence, this introduces a region of hysteresis whereby a single count is made only when the signal drops below the maximum threshold and emerges below the minimum threshold, or vice versa. This is called threshold-crossing rate (TCR), and is illustrated in Figure 6.3.

In practice, the advantage of TCR for noise reduction is often achieved by low-pass filtering the speech before a ZCR is calculated. This knocks out the high frequency noise or 'bounce' on the signal. Since ZCR is used as a rough approximation of the fundamental pitch of an audio signal, bounds for filtering can be established through knowing the extent of the expected maximum. In speech, it has to be stressed that the filtered ZCR (or TCR) measure provides an approximate indication of the content of the speech signal, with unvoiced speech tending to result in high ZCR values, and voiced speech tending to result in low ZCR values. Noise also tends to produce high ZCR values, and thus it is difficult to use ZCR for analysis of noisy speech, significantly limiting its use in practical applications.

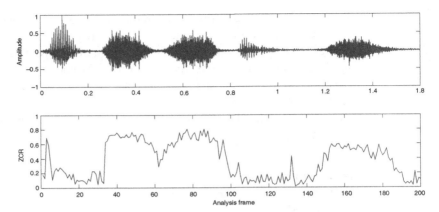

**Figure 6.2** A 16 kHz sampled speech recording 'its speech', containing several stressed high frequency sibilants (plotted in the upper graph), and the ZCR measure corresponding to this (lower graph). The excursions of the ZCR plot follow the high frequency speech components, namely a breathy /i/ at the start, both /s/ sounds, and the final /ch/.

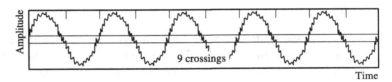

**Figure 6.3** Threshold-crossing rate illustrated for a noisy sinewave, showing that the extra zero crossings of Figure 6.1(b) are no longer present.

In most ways, TCR results are very similar to ZCR. In the absence of noise, a TCR plot for a speech recording would resemble that for ZCR as in Figure 6.2.

## 6.1.2    Frame power

This is a measure of the signal energy over an analysis frame, and is calculated as the sum of the squared magnitude of the sample values in that frame. For speech frame $i$, with $N$ elements, denoted by $x_i(\ )$, the frame power measure is determined from:

$$E_i = \frac{1}{N} \sum_{n=0}^{N-1} |x_i(n)|^2 . \tag{6.2}$$

As with the case of ZCR in Section 6.1.1, the division by $N$ will often be unnecessary in practice.

In MATLAB, that is a very simple formula:

```
function [fpow]=fpow(segment)
    fpow=sum(segment.^2)/length(segment);
```

Frame power provides a compact representation of the amplitude of the speech. As we have seen in Section 3.2.3, unvoiced speech is spoken with less power than voiced speech, and for this reason, frame power provides another indicator of voiced/unvoiced speech.

The simplicity of the MATLAB function above hides the fact that this calculation requires a multiplication to be performed for each and every sample within the analysis window. In implementation terms this can be relatively 'expensive', prompting simplification efforts which led directly to the AMDF below. In fact the similarity of the two measures is illustrated in Figure 6.4, which plots the frame power and AMDF measures together for an example recording of the first seven letters of the alphabet. The speech was recorded with a 16 kHz sample rate, and analysis performed on non-overlapping 128-sample frames. Each of the plots is scaled to a maximum of 1.0 for comparison purposes.

## 6.1.3      Average magnitude difference function

The average magnitude difference function is designed to provide much of the information of the frame power measure, but without multiplications:

$$\text{AMDF}_i = \frac{1}{N} \sum_{n=0}^{N-1} |x_i(n)|. \tag{6.3}$$

In MATLAB, it is again very simple:

```
function [amdf]=amdf(segment)
    amdf=sum(abs(segment))/length(segment);
```

An illustration of AMDF obtained for a sequence of recorded speech is shown in Figure 6.4. This also plots the frame power measure, and illustrates the quite close correspondence of the two measures. Both output high values when speech power is high (such as the /a/ sound in the letter A and both the /c/ and /ee/ sounds of the letter C) and output low measure results when speech power is low.

Although correspondence between frame power and AMDF appears to be quite close in this plot, it should be noted that the AMDF output is higher than frame power during the gaps between words. This is an indicator that the AMDF may be less immune to confusion by noise than the frame power measure.

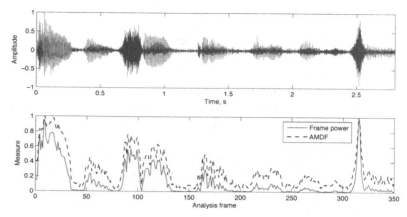

**Figure 6.4** Average magnitude difference function (AMDF) and frame power plots (lower graph) for 16 kHz sampled speech containing a recitation of the alphabet from letter A to letter G (plotted on the upper graph). The duration of the letter C is enough for there to be almost two amplitude 'bumps' for the /c/ and the /ee/ sounds separated by a short gap, spanning the time period from approximately 0.7 to 1.2 seconds. Note that the correspondence of the two analysis measures plotted on the lower graph is close during periods of speech, but less so in the gaps between words.

## 6.1.4    Spectral measures

In Chapter 2 we had looked at how to use an FFT to determine a frequency spectrum. In MATLAB we had created and then plotted the spectrum of a random vector (see Section 2.3). If, instead of plotting the spectrum directly, we were to analyse the spectral components, we could use these to build up a spectral measure.

To illustrate this, we will plot the spectra of two different regions of a speech recording, examine these and analyse further. Both the time and frequency domain plots are shown in Figure 6.5, for the spoken letters C and R.

From the time-domain waveform plots in the figure, it can be seen that the C is probably louder than the R (it has higher amplitude), and also has a slightly longer duration. The frequency-domain plots show some spikes at low frequencies – most likely formant frequencies – but also show that the C has more high-frequency components than the R. Most of the signal power in the R seems to be below about 1 kHz, but much of the signal power in the C seems to be above this.

In MATLAB we can devise a measure to compare the low-frequency spectral power to the high-frequency spectral power. To do this, first we need to derive the spectra (as plotted in the lower half of Figure 6.5) and name them `fft_c` and `fft_r` to denote the spectra of spoken letter C and spoken letter R, respectively:

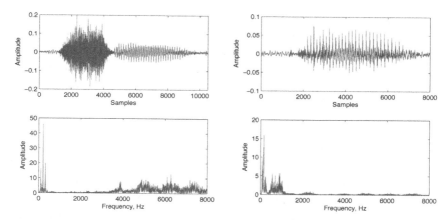

**Figure 6.5** Plot of two 16 kHz sampled speech utterances of the letters C (left) and R (right), with time-domain waveform plots at the top, and frequency-domain spectra plotted below them.

```
Nc=length(speech_letter_c);
Nr=length(speech_letter_r);
fft_c=fft(speech_letter_c);
fft_c=abs(fft_c(1:Nc/2));
fft_r=fft(speech_letter_r);
fft_r=abs(fft_r(1:Nr/2));
```

At this point, we could plot the spectra if required. Remember, we are plotting only the positive frequency components: the FFT output would be as long as the original speech array, so we take the lower half of this only for simplicity, and we take the absolute value because the FFT output is complex. When trying this example, MATLAB may complain that you need to ensure that only integers are used for the 'colon operator' indexing. This would happen if the length of the original speech arrays was not even (in fact, considering the further subdivision we are about to perform, we should really ensure that the speech array sizes were a multiple of four).

Next we can simply sum up the frequency elements within the required ranges (in this case, the lower half frequencies and the upper half frequencies respectively):

```
c_lowf=sum(fft_c(1:Nc/4))/(Nc/4);
c_highf=sum(fft_c(1+Nc/4:Nc/2))/(Nc/4);
r_lowf=sum(fft_r(1:Nr/4))/(Nr/4);
r_highf=sum(fft_r(1+Nr/4:Nr/2))/(Nr/4);
```

For the example spectra plotted in Figure 6.5, the results are telling. Splitting the spectrum in half, the mean absolute lower half frequency components for R are 0.74, and for C are 0.87. For the mean absolute higher half frequency components, R scores 0.13

while C scores 2.84. However it is the ratios of these that are particularly meaningful. The letter C has a high-frequency to low-frequency ratio of 3.3, but the letter R scores only 0.18. These figures indicate that much of the energy in the spoken letter C is higher frequency, whilst much of the energy in the spoken letter R is lower frequency. Indeed we can visualise this by looking at the spectral plots, but we have just created a measure that can be performed automatically by a computer:

```
c_ratio=c_highf/c_lowf;
r_ratio=r_highf/r_lowf;
```

Although this is a relatively trivial example, it is possible, for one speaker, to identify spoken letters by segmenting them (isolating an analysis frame that contains a single letter, and even subdividing this), performing an FFT, then examining the ratio of the summed frequency components across different regions of the frequency spectrum. Unfortunately this technique cannot normally be generalised to work for the speech of many different people.

## 6.1.5    Cepstral analysis

The cepstrum was introduced in Section 2.6.2.2 where an example of the technique was presented as a useful method of visualising speech signals. As with many other visualisation methods, the useful information that the human eye can notice in a plot can also be extracted and analysed by computer.

The usefulness of the cepstrum derives from the fact that it is the inverse FFT of the logarithm of the FFT. In general terms, this means that the frequency components have been ordered logarithmically. In mathematics, one of the principles of the logarithm is that if something is the multiplicative combination of two items, then in the logarithmic domain, these items are combined additively. Put another way, if a signal under analysis, $y(t)$ can be said to be equal to $h(t)$ multiplied by $x(t)$, then:

$$y(t) = h(t) \times x(t) \tag{6.4}$$

$$\log[y(t)] = \log[h(t)] + \log[x(t)].$$

Relating back to speech signals, $x(t)$ may well be a pitch component, while $h(t)$ is a vocal tract component. In the time domain these are related multiplicatively, but in the cepstrum domain, they are related additively. In a cepstral plot then, the pitch component, for instance, would be visible in its own right, separated from the vocal tract component.

This has been illustrated in Figure 6.6, plotted using the method of Section 2.6.2.2. The most likely position of the fundamental pitch period component, at index position 64, has been selected.

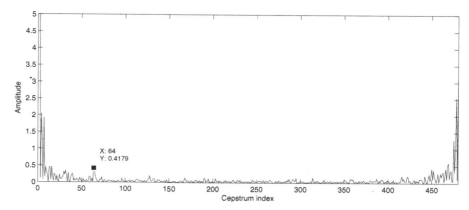

**Figure 6.6** Cepstral plot of a segment of voiced speech, amplitude against cepstral index for a 480-sample analysis window. The likely pitch component has been selected, at index of 64.

Cepstral analysis is also used for many other purposes than pitch detection. One of the more useful, and traditional, uses is in the extraction of spectral envelope information for speech analysis. In general, the spectral envelope is the smoothed shape of the frequency plot, where the smoothing process really means ignoring the higher frequency components. In the cepstral domain, this can be performed by discarding all cepstral coefficients related to frequencies higher than the envelope frequency.

If this is performed, and then the FFT plotted, the smoothing process can be quite obvious. MATLAB code is provided below which performs this process, and illustrated, acting on a 480-sample window of voiced speech, named segment, in Figure 6.7.

```
len=length(segment);
%Take the cepstrum
ps=log(abs(fft(segment)));
cep=ifft(ps);
%Perform the filtering
cut=30;
cep2=zeros(1,len);
cep2(1:cut-1)=cep(1:cut-1)*2;
cep2(1)=cep(1);
cep2(cut)=cep(cut);
%Convert to frequency domain
env=real(fft(cep2));
act=real(fft(cep));
%Plot the result
pl1=20*log10(env(1:len/2));
pl2=20*log10(act(1:len/2));
span=[1:fs/len:fs/2];
plot(span,pl1,'k-.',span,pl2,'b');
xlabel('Frequency, Hz');
```

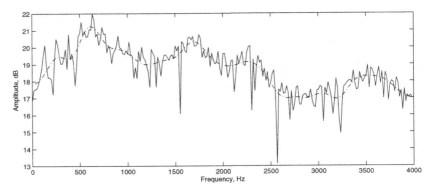

**Figure 6.7** Frequency plot of a segment of voiced speech (solid line) overlaid with the frequency envelope obtained from the first few cepstral coefficients (dashed line).

```
ylabel('Amplitude, dB');
```

It will be noticed in the MATLAB code, that as well as discarding the unwanted cepstral coefficients in the cep2 array [1], all of the components between DC and the one immediately below the cutoff frequency are accentuated by being doubled, to better match the required envelope.

Apart from performing the cepstrum manually, MATLAB contains much faster functions to calculate the cepstrum and its inverse in the signal processing toolbox, namely cceps and icceps. These could have been used synonymously with the operations in the MATLAB code above.

## 6.1.6    LSP-based measures

Analysis can also be performed using LSP-based measures since usable and interpretable information can be extracted from raw LSPs. These analysis techniques are normally used for speech, although to demonstrate their wider application, examples will also be given illustrating the analysis of a musical instrument and of an animal sound.

For the following sections, it is assumed that LSP analysis, as described in Section 5.2.4, has been performed, and that a set of line spectral pairs is available for an analysis window. We will first consider instantaneous LSP measures which are means of interpreting a single set of lines from one analysis frame, before we consider the longer term trends of these measures.

### 6.1.6.1     Instantaneous LSP analysis

Although there are many possible methods of analysis, in this section we will consider just three features which can be used for individual analysis frames of speech. Later, in Section 6.1.6.2, we will then extend the analysis to the evolution of the measures through time. In the following equations, subscript $i$ refers to one of the $p$ LSPs $\omega_i$ which represent speech in analysis frame $n$.

The overall shift in LSP location from one frame to the next, referred to as the gross shift, *Shift*, indicates predominant spectral movements between frames:

$$Shift[n] = \left\{ \sum_{i=1}^{p} \omega_i[n] \right\} - \left\{ \sum_{i=1}^{p} \omega_i[n+1] \right\}. \tag{6.5}$$

In MATLAB this would be:

```
function [shift] = lsp_shift(w1,w2)
    shift=sum(w1) - sum(w2);
```

The mean frequency of all LSPs representing an analysis frame is also a useful measure of frequency bias, *Bias*:

$$Bias[n] = \frac{1}{p} \sum_{i=1}^{p} \omega_i[n] \tag{6.6}$$

or in MATLAB:

```
function [bias] = lsp_bias(w)
    bias=sum(w)/length(w);
```

It is also possible to specify a nominal LSP positioning, and calculate a deviation *Dev* between this reference LSP positioning $\bar{\omega}_i$ and the LSPs under analysis, raised to an arbitrary power $\beta$. This is measured as follows:

$$Dev[n] = \sum_{i=1}^{p} (\omega_i[n] - \bar{\omega}_i)^{\beta}. \tag{6.7}$$

A suitable MATLAB function for calculating the deviation is shown below:

```
function [dev] = lsp_dev(w,bar_w,b)
    dev=sum( (w-bar_w).^b);
```

where $\bar{\omega}_i$ would often be a set of nominally positioned LSPs equally spaced across the frequency domain:

$$\bar{\omega}_i = i\pi/(p+1) \qquad \text{for} \quad i = 1, \ldots, p. \tag{6.8}$$

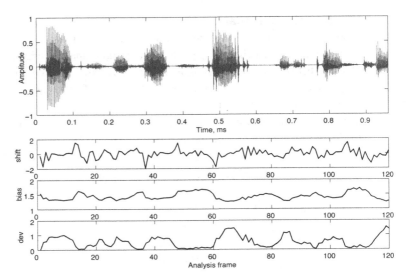

**Figure 6.8** Deviation, bias and shift LSP analysis features collected for a 16 kHz sampled speech utterance, with waveform plotted on top, extracted from the TIMIT database.

In MATLAB, this is again easy, assuming the $p = 10$ we would simply create the reference LSPs using:

```
bar_w=[1:10]*pi/11
```

With the $p$ lines distributed evenly across the spectrum, if this were transformed into LPC coefficients and their power spectrum calculated, it would be flat. *Dev* thus determines how close each frame is to this distribution, such that with $\beta = 2$, it becomes the Euclidean distance between the actual and comparison distributions. Odd values such as $\beta = 1$ or $\beta = 3$ attribute a sign to the deviation from $\bar{\omega}_i$, so that a positive measure denotes high-frequency spectral bias, and a negative measure specifies a low-frequency spectral bias.

Each of these measures provides useful information regarding the underlying speech signal, and are illustrated when applied to a speech recording from the TIMIT database [2], in Figure 6.8 (the deviation plot is given for $\beta = 2$).

*Shift* indicates predominant LSP frequency distribution movements between consecutive frames. Considering an example of two adjacent frames containing unvoiced and voiced speech, the LSP distributions in the two frames will be low-frequency, and high-frequency biased respectively. We saw a similar effect when comparing the spoken C and R spectra in Section 6.1.4. A large difference between the two frames gives a large measure value. The shift measure as shown in Figure 6.8, peaks at obvious speech waveform changes, and may thus be advantageous for speech segmentation.

*Bias* indicates frequency trends within the current frame – that is whether the spectrum of the current frame is high-frequency or low-frequency biased. It is similar to the deviation measure which determines how close the LSP distribution of the current frame is to a predetermined comparison distribution. In Figure 6.8 this registers high values for fricatives, indicating the predominance of their high-frequency components.

Where the speech is contaminated by noise of a particularly well-defined shape, if the comparison distribution, $\bar{\omega}$, of Equation (6.7) is set to represent a spectrum of this shape, then the *Dev* measure may be reasonably insensitive to noise when averaged. In other words analysing the noise itself will produce a zero mean output.

We can also use LSP data to estimate the position of spectral peaks within an analysis frame. Peaks are located approximately halfway between pairs of closely spaced lines, with the peak power related to the closeness of the lines. A ranking of the first few most closely-spaced line pairs in terms of narrowness, will generally correspond to the ordering of the corresponding peaks by power. It must be noted, however, that in some cases, especially where three lines are close together, the correspondence is far less predicable. For unvoiced frames, other speech resonances are similarly reflected in the LSP distribution, although the visual correspondence when plotted is far less dramatic than in the case of strongly voiced speech.

### 6.1.6.2    Time-evolved LSP analysis

Each of the instantaneous measures can be tracked to build up a statistical model of the signal under analysis. Similarly, spectral peaks detected by a measure of line-closeness, can be tracked and post-processed to extract likely formant positions. Analysis of LSP feature changes over time may be combined with consideration of traditional speech features to track dynamic speech characteristics.

Applications of speech feature tracking can include the classification of speech analysis frames into voiced, unvoiced or non-speech periods. Evidence suggests that instantaneous speech features derived from LSP analysis are similar, under low noise conditions, to zero-crossing rate [3].

A simple example of speech classification uses a binary LSP decision measure of upward or downward deviation. This LSP vote measure counts the number of lines in a frame above their median value, and is a computationally simple alternative to Equation (6.7). This has been used by the author to perform phonetic classification of speech [3]. In this application the LSP vote measure was used to categorise speech frames into six classes: voiceless fricatives and affricatives, fricatives, plosives, nasals and glides, and silence.

At this point we should note that by far the most popular application of continuous LSP analysis is in the speech recognition field. In general, speech recognition uses techniques such as template matching, statistical modelling or multi-feature vector quantisation. It is possible to use LSP-derived data with each of these techniques, although the statistical matching of a hidden Markov model (HMM) appears to be a more common application for LSP-derived features [4].

Finally, continuous LSP analysis has been used to collect statistics on the line frequency distribution over time in a CELP coder and relate this to the language being conveyed. It has been shown that LSP statistics differ markedly across several languages [5], leading to the potential for LSP-based features to be used in linguistic detection algorithms.

## 6.2    Speech analysis and classification

The analysis of speech is an important requirement of many different applications and the classification of speech into various categories is a necessary part of many techniques. A full list would be lengthy, but the following subset of basic techniques indicates the sheer range of applications of speech analysis and classification:

- detecting the presence of speech;
- detecting voiced or unvoiced speech;
- finding boundaries between phonemes or words;
- classifying speech by phoneme type;
- language detection;
- speaker recognition;
- speech recognition.

Classification is an important, and growing area of speech research which relates to the machine 'understanding' of speech (where understanding can range from knowing whether speech is present right through to understanding the meaning or emotion conveyed by the spoken word).

In order to begin such classification of speech, it is usually necessary to first perform some form of measurement on the speech signal itself. For example detecting voiced or unvoiced speech might require the determination of speech power and pitch, perhaps through examination of LSP data. Many methods can potentially be used for analysis of speech, and extensive empirical testing is almost always required to determine the best subset of measures to be used for a particular application, whatever that may be.

By and large, new classification tasks will make use of many of the same basic speech measures as described in our analysis toolkit (Section 6.1). Measures can be used in different ways, to different ends, and most often several measures must be combined together.

In addition to the several methods in our toolkit, there are some methods which have been very much reserved for speech analysis, particularly pitch detection-related methods, since pitch is so important to speech communications.

In the following subsections, we will examine techniques for the analysis and extraction of pitch from speech.

## 6.2.1     Pitch analysis

Section 5.3.2.1 discussed the use of long-term prediction to determine the pitch of a speech waveform, a method commonly used in speech coders. There are, however, many alternatives. The most accurate require some form of human involvement, or a measurement device such as an accelerometer or electroglottograph (EGG) on the throat or glottis producing the pitch. From a recorded speech waveform alone, it is currently impossible to identify a definitive answer to the question 'what is the pitch?', but nevertheless some algorithms appear to get close. Tests of these pitch analysis algorithms tend to rely on consensus answers. As an example, many methods have been surveyed, and compared, in an excellent paper by Rabiner *et al.* [6].

Some of the more mainstream techniques reported by Rabiner and others operating purely on recorded sound include the following:

- time-domain zero-crossing analysis (perhaps thresholded – see Section 6.1.1);
- time-domain autocorrelation (the method used in Section 5.3.2.1);
- frequency-domain cepstral analysis (see Section 2.6.2.2);
- average magnitude difference function based methods (see Section 6.1.3);
- simplified inverse filtering technique (SIFT);
- LPC and LSP-based methods (such as those in Section 6.1.6);
- time-frequency domain analysis – explained in Section 6.2.2.

We will not reproduce the excellent work of Rabiner here, but will introduce a modern alternative, the use of time-frequency distribution analysis. Whilst being more complex than most, if not all, of the other methods mentioned, early indications are that this method of analysis is promising in terms of achieving better accuracy than the traditional techniques.

## 6.2.2     Joint time-frequency distribution

Joint Time-Frequency Distribution (TFD) analysis originally emerged in the radar signal processing field, but has started to be adopted for speech processing in recent years [7]. Its good performance in tracing frequency transitions as time progresses has been noted by speech researchers. Given the importance of pitch in speech systems, it is thus little surprise that TFD analysis has been attempted for pitch determination.

We will discuss four joint time-frequency distributions, as described in [8], for use in pitch determination. These are namely the Spectrogram Time-Frequency Distribution (STFD), Wigner–Ville Distribution (WVD), Pseudo-Wigner–Ville Distribution (PWVD) and Reassigned Smoothed Pseudo-Wigner–Ville Distribution (RSPWVD). They all attempt to identify the characteristics of frequency as it changes with time from slightly different points of view. Each one of these TFD algorithms has its own strengths and weaknesses relating to the sensitivity to detect pitch features and the implementation complexity (and these have been well established in research literature).

The STFD computes the spectrogram distribution of a discrete-time signal $x$. It corresponds to the squared modulus of the short-time Fourier transform (see Section 2.6).

As such it is quite easily described as an integral across analysis window $w$ in Equation (6.9).

$$S_x(t, v) = \left| \int x(u)w(u - t)e^{-j2\pi vu} du \right|^2 .$$ (6.9)

The instantaneous frequency energy at time $t$ is then given by:

$$E_x(t) = \frac{1}{2} \int S_x(t, v)dv.$$ (6.10)

The Wigner–Ville distribution (WVD) is a bilinear distribution that is qualitatively different from the STFD. At any particular time, the signal contributions from the past are multiplied by the signal contributions from the future in order to compute and correlate between these left and right parts of the signal. It has the ability to give a clearer picture of the instantaneous frequency and group delay than the spectrogram:

$$WV_x(t, v) = \int x\left(t + \frac{\tau}{2}\right) x\left(t - \frac{\tau}{2}\right) e^{-j2\pi v\tau} d\tau.$$ (6.11)

Similarly, the instantaneous frequency energy at time $t$ is then given by:

$$E_x(t) = \frac{1}{2} \int |WV_x(t, v)| \, dv.$$ (6.12)

Although the WVD is perfectly localised on linear signals due to the forward-backward nature of the analysis. However, if there are several frequency components existing simultaneously, noise will be found in the WVD distribution due to the phenomenon of cross-term interference between those components. It also provides equal weighting to both past and future components.

The pseudo-Wigner–Ville distribution (PWVD) is an advance on the WVD since it emphasises the signal properties near the time of interest compared to far away times. A window function $h(\tau)$ is used that peaks around $\tau = 0$ to weight the Wigner–Ville distribution toward an emphasis of the signal around time $t$. The PWVD is defined as:

$$PWV_x(t, v) = \int h(\tau)x\left(t + \frac{\tau}{2}\right) x\left(t - \frac{\tau}{2}\right) e^{-j2\pi v\tau} d\tau.$$ (6.13)

Again, the instantaneous frequency energy calculation at time $t$ is relatively simple:

$$E_x(t) = \frac{1}{2} \int |PWV_x(t, v)| \, dv.$$ (6.14)

Finally, a fourth method exists as a refinement to PWVD. This is the reassigned smoothed PWVD (refer to [8] for its application). It uses a centre-of-gravity reassignment method with both time and frequency domain smoothing.

These methods have each been tested within a GSM structure for the determination of pitch [8]. The time-frequency distribution algorithms, with a little post-processing, were used as a replacement for the original pitch analysis used for the RPE structure. This overall system was then tested to determine what changes (if any) resulted in the intelligibility of speech conveyed. In each case the same speech and other conditions were used for the tests, and the Chinese diagnostic rhyme test (CDRT) – a Chinese language equivalent of the diagnostic rhyme test (DRT) of Section 3.3.3 – was used to assess intelligibility. Results indicated a significant improvement in intelligibility for the speech that differed in terms of sibilation, and smaller less significant improvements in several other speech classes.

Overall it seems that the results are promising; TFD might well become another candidate for pitch determination in speech communications systems. However at present other methods, particularly the autocorrelation based systems, are most popular.

## 6.3     Analysis of other signals

Arguably the predominant application of audio analysis has been related to speech communications applications, with a notable (and economically important) extension into music compression, such as MP3. However there is no reason why the techniques, and toolkit introduced in this chapter should not be applied elsewhere. As an example, one of the early algorithms produced by the author was capable of detecting and classifying dog barks, and in 1997 the author was contacted by a team of economists who wished to apply LSP analysis to the stock market.[1]

### 6.3.1     Analysis of music

Many researchers have investigated musical instrument recognition and coding. For example, Krishna and Sreenivas [9] evaluated three methods of recognising individual instrument sets. One of these methods was based upon LSPs and was found to be the superior method among those tested. The authors stated several advantages of LSPs including their localised spectral sensitivities, their ability to characterise both resonance locations and bandwidths (characteristics of the timbre of the instruments), and the important aspect of spectral peak location.

In order to examine some of these claims, it is easy to use MATLAB for analysis. As an example, MATLAB was used to record a violin open A string (tuned to 440 Hz), sampled at 16 kHz.

Those readers with a violin handy may wish to record a short segment using the methods of Chapter 2. It is important to ensure that the player uses an uninterrupted and smooth bowing action during the recording. Given such a recording in a floating point

---

[1] Sadly nothing was subsequently heard from the economists, from which we can conclude that either the technique did not work, or was so successful that they immediately retired from research before writing up, and did not wish to share their key to unlimited wealth.

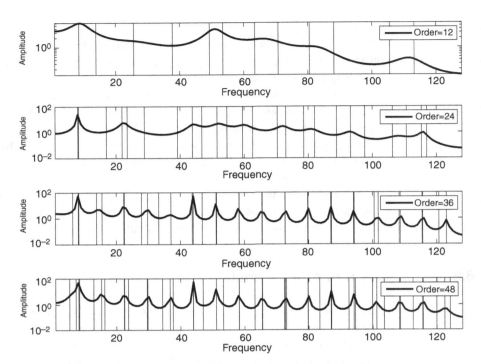

**Figure 6.9** LSP locations and LPC power spectrum of a violin open A string plotted for various orders of analysis, against a frequency index from 1 to 128.

vector named `violin`, we can perform a $P$th-order LPC analysis, convert to LSPs and plot the result.

```
P=16;
Ws=256;   %window size=256 samples
vseg=violin(1:Ws).*hamming(Ws);
a=lpc(vseg, P);
w=lpc_lsp(a);
lpcsp(a, w);
```

There should be a very obvious spectral peak located at 440 Hz, with several harmonics visible. Figure 6.9 shows a plot of this analysis, ranging from 12th to 48th order, revealing greater and greater levels of spectral detail as the order increases. In this case the analysis shows a 4096-sample section. In each case there is a reasonably good correspondence of the LSP positions to spectral peaks – in particular the location of narrow line pairs around many major peaks. As more detail is revealed through increased order analysis, the harmonic structure of the played note becomes clearer, although a triple (rather than pair) at the fundamental of the 48th-order plot probably indicates that the analysis order is greater than the note complexity warrants.

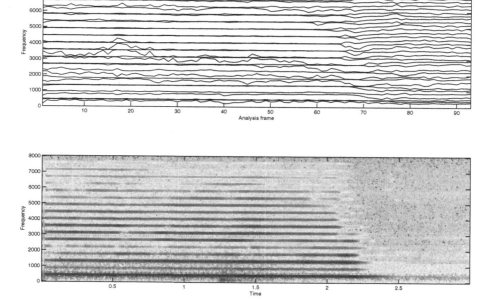

**Figure 6.10** LSP tracks for a violin note played on the open A string sampled at 16 kHz (top) compared to a spectrogram of the same recording (bottom).

Another plot shows the note as it was played, analysed through a time evolution. This is shown in Figure 6.10, where the upper graph displays the 36th-order LSP values as they evolve over time (note that although it looks as if they may do, the lines do not cross, since they are always monotonically ordered), for each 512-sample analysis window. Below this, a spectrogram of the same recording is plotted, using a 256-sample window size with 50% overlap. Note the LSP narrowness around the obvious 440 Hz fundamental and harmonics shown in the spectrogram. At the end of the played note, some resonance continues to sound a 440 Hz fundamental but with only weak harmonics (once the abrasion of the rosin-coated horsehair bow on the aluminium-clad string has ceased). During this decay period the upper LSPs gradually lose their tendency to pair up and begin to flatten out. However the fundamental continues to be marked by an ongoing narrow spacing of the lower two lines.

The close visual correspondence between the spectrogram and the LSP evolution plot in Figure 6.10 supports the assertion of Krishna and Sreenivas [9] mentioned previously that LSP values are usable for instrument recognition. Most likely both fundamental and harmonic frequency identification can be determined through automatic analysis of LSP data, especially when combined with other measures such as ZCR or frame power.

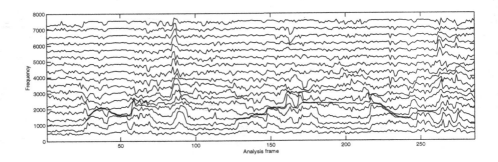

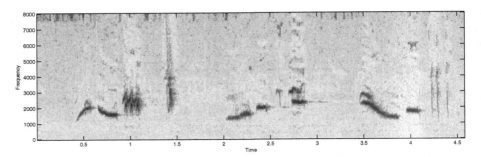

**Figure 6.11** LSP tracks for a few seconds of a blackbird's song sampled at 16 kHz (top) compared to a spectrogram of the same recording (bottom).

### 6.3.2    Analysis of animal noises

A recording of birdsong from the BBC was analysed in a similar fashion to the analysis of the violin note in Section 6.3.1 [10]. After some experimentation, an 18th-order LPC analysis was found to be optimal for LSP track plotting as shown in Figure 6.11.

The analysis covers approximately 4.5 seconds of a blackbird's warble, and shows a clear visual correlation between the birdsong syllables shown in the spectrogram, and in the plotted LSP trajectories. The spectrogram had 50% overlapped windows of size 256 samples, and non-overlapped 256-sample windows were used for the LSP tracking. Other authors report that harmonic analysis has been shown useful in the classification of birdsong [11], and since both plots in Figure 6.11 so clearly show harmonic effects (look for LSP pairs shifting together), the potential for such analysis using these methods is clear.

As a final analysis of animal noise, a recording of a very large and angry dog barking was obtained, at some risk, using a 16 kHz sample rate. Again this was analysed by first obtaining LPC coefficients, and then a set of LSPs, for a succession of 64-sample analysis frames. The analysis order was set to eight in this instance since the pitch of the dog bark was rather low, and strongly resonant. Figure 6.12 plots the time-domain waveform above the time-evolution plot of the eight LSP tracks. A bar graph overlaid upon this depicts frame power, and both power and LSP frequencies were scaled to be between 0 and 1 for ease of plotting.

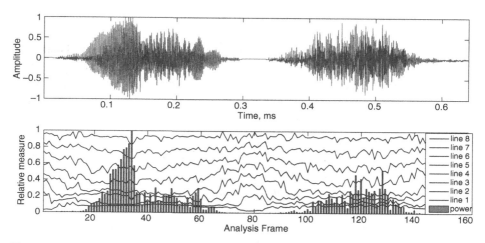

**Figure 6.12** Waveform plot (top) and LSP tracks (below) with overlaid frame power plot (bars on lower graph) for a double dog bark, recorded at 16 kHz and an analysis order of eight.

The dog bark waveform of Figure 6.12 resembles human speech – in fact a recording of a human shouting 'woof woof' may look similar. The narrow pairs of LSPs, observable during the periods of highest power, indicate the resonance of strong formants during the loudest part of the bark. The similarity with human speech would indicate that methods of analysis (and processing) used for human speech could well be applicable to animal noises. In particular, vocally generated animal noises, using the same mechanisms as human speech, would be more amenable to speech processing methods than those produced in other ways, such as the abrasive sounds of grasshoppers, dolphin squeaks, pig snorts, snake rattles, and so on.

## 6.4      Higher order statistics

When applied to speech, the measures described in this section do not provide a definitive indication of the presence or absence of speech. Generally a definitive classification result is only practical under certain tightly controlled conditions (such as same user, same microphone, small vocabulary, etc.). In the real world, things become more difficult than in the lab. The author can recall undertaking a very lucrative contract to develop a commercial speech classification engine over several months. All went well initially: algorithms were implemented, background noises and speech samples collected, and a working system rapidly developed. However, problems arose during testing: one person, for no apparent reason, had the ability to 'break' the system every time with his voice. Despite little discernible difference to other speakers, this voice confused the combined measures used for classification. A large amount of adjustment and rework was required to modify the system to correct for this. The moral of this story is that, even developing with a large number of representative samples of speech and background noise, the sheer variability of voices and sounds in the real world spells trouble for the speech processing engineer.

Speech classification is a matter of statistics. A system which classifies unconstrained speech 100% correctly is impossible. However scores approaching this are feasible. Most important is where and when any system goes wrong. The developer needs to bear in mind four conditions regarding the accuracy of a binary classification (positive-negative match to some criteria):

- True-Positive classification accuracy
  *the proportion of positive matches classified correctly*;
- True-Negative classification accuracy
  *the proportion of negative matches classified correctly*;
- False-Positive classification accuracy
  *the proportion of negative matches incorrectly classified as positive*;
- False-Negative classification accuracy
  *the proportion of positive matches incorrectly classified as negative*.

In general, accuracy can be improved by ensuring a larger sample size. However the way that this is analysed is important. For example, although we can say that speech occupies certain frequency ranges, at certain volumes, there are many types of sound that are similar. Dog growls, chairs scraping across floors and doors slamming could appear like speech if spectral measures alone are used. By and large, we would therefore need to use more than one basic measure – perhaps spectral distribution and amplitude distribution (AMDF measure). Unfortunately, those measures would be confused by music – similar frequency range to speech, and similar amplitude changes. There are thus other aspects that we need to look for.

Specifically for speech, there are several higher order statistics that we can turn to, and which are seldom present in generalised audio. These relate to the usage, generation, and content of the speech signal itself.

First is the *occupancy rate* of the channel. Most speakers do not utter long continuous monologues. For telephone systems, there are generally pauses in one person's speech, which become occupied by the other party. These to-and-fro flows can be detected, and used to indicate the presence of speech. For this, however we may require analysis of several minutes' worth of speech before a picture emerges of occupancy rate.

The *syllabic rate* is the speed at which syllables are formed and uttered. To some extent, this is a function of language and speaker – for example, native Indian speakers have a far higher syllabic rate than native Maori speakers, and irrespective of origin, most people do not exhibit high syllabic rate when woken in the early hours of the morning. However, the vocal production mechanisms are muscle-controlled, and are only capable of a certain range of syllabic rate. This can be detected, and the fact used to classify speech.

Most languages have a certain *ratio of voiced and unvoiced* speech. Degree of voicing can be affected by several conditions such as sore throat, and speaking environment (think of speaking on a cellphone in a library), but there is still a pattern of voiced and unvoiced speech in many languages. In Chinese, all words are either totally voiced (V),

Voiced-Unvoiced (VU), UV or VUV. In English the pattern is far more random, but there *is* a pattern, and given sufficient raw material, it can be detected and used for classification.

Finally the speech *cadence* is similar to syllabic rate but is effectively the rate at which words are spoken. Obviously this depends upon many factors, not least the length of words being spoken, however it maintains a surprisingly narrow range of about 2 Hz to 7 Hz for normal speech. This can be determined from the frequency of gaps between words. But this is not as easy as it may appear since the gaps themselves are often extremely difficult to detect automatically, and often practically non-existent. Some word patterns have longer gaps inside words than between them. We will touch upon some of these issues once again when we consider some different types of speech classification in Chapter 7.

## 6.5     Summary

This chapter described the 'analysis toolkit' of standard techniques available to the audio engineer, along with the methods of integrating these techniques into an analysis system. The important aspect of pitch extraction for speech analysis was also discussed with reference to several useful techniques.

Apart from speech, examples were given of some of these techniques applied to other non-speech audio signals, namely a musical instrument recording and animal noises.

Next, in Chapter 7, several of the analysis and processing techniques presented to date will be integrated and used in more complex and advanced speech or audio processing algorithms.

## Bibliography

- *Acoustic Analysis of Speech*
  R. D. Kent (Singular, 2nd edition 2001)

- *Speech and Audio Signal Processing: Processing and Perception of Speech and Music*
  B. Gold and N. Morgan (Wiley, 1999)

- *Advances in Speech, Hearing and Language Processing*
  Ed. W. A. Ainsworth (JAI Press, 1990)

- *Applications of Digital Signal Processing to Audio and Acoustics*
  Ed. M. Kahrs (Springer, 1998)

- *Music: A Mathematical Offering*
  D. Benson (Cambridge University Press, 2006)
  This book is a fascinating study of the close relationship between mathematics and music. Although not describing the practical computational analysis of sound or music, the theoretical background and methods of analysis and science of music and of musical instruments are well described. This book starts with the human ear, and progresses beyond the Fourier transform to music technology.

# References

[1]   L. R. Rabiner and R. W. Schaefer. *Digital Processing of Speech Signals*. Prentice-Hall, 1978.

[2]   TI and MIT TIMIT database. *A CD-ROM database of phonetically classified recordings of sentences spoken by a number of different male and female speakers*, disc 1-1.1, 1990.

[3]   I. V. McLoughlin. LSP parameter interpretation for speech classification. In *Proceedings of the 6th IEEE International Conference on Electronics, Circuits and Systems*, number 113, September 1999.

[4]   K. K. Paliwal. A study of LSF representation for speaker-dependent and speaker-independent HMM-based speech recognition systems. In *Proc. Int. Conf. on Acoustics, Speech and Signal Processing*, Volume 2, pages 801–804, 1990.

[5]   J. Parry, I. Burnett, and J. Chicharo. Linguistic mapping in LSF space for low-bit rate coding. *Proc. Int. Conf. on Acoustics, Speech and Signal Processing*, Volume 2, pages 653–656, March 1999.

[6]   L. R. Rabiner, M. Cheng, A. Rosenberg, and C. McGonegal. A comparative performance study of several pitch detection algorithms. *IEEE Trans. Acoustics, Speech Signal Proc.*, **24**(5): 399–418, 1976.

[7]   L. Cohen. *Time-Frequency Analysis*. Prentice-Hall, 1995.

[8]   Z. Q. Ding, I. V. McLoughlin, and E. C. Tan. How to track pitch pulse in LP residual – joint time-frequency distribution approach. *In Proceedings of the IEEE Pacific Rim Conference on Communications, Computers and Signal Processing*, August 2001.

[9]   A. G. Krishna and T. V. Sreenivas. Musical instrument recognition: From isolated notes to solo phrases. In *IEEE International Conference on Acoustics, Speech and Signal Proc.*, Volume 4, pages 265–268, 2004.

[10]  The British Broadcasting Corporation BBC. BBC Radio 4: Brett Westwood's guide to garden birdsong, May 2007. URL http://www.bbc.co.uk/radio4/science/birdsong.shtml.

[11]  A. Harma and P. Somervuo. Classification of the harmonic structure in bird vocalization. In *IEEE International Conference on Acoustics, Speech and Signal Proc.*, Volume 5, pages 701–704, 2004.

# 7    Advanced topics

So far this book has dealt with individual subjects ranging from background material on audio, its handling with MATLAB, speech, hearing, and on to the commercially important topics of communications and analysis. Each of these has been relatively self-contained in scope, although the more advanced speech compression methods of the previous chapter did introduce some limited psychoacoustic features.

In this chapter we will progress onward: we will discuss and describe advanced topics that combine processing elements relating to both speech and hearing – computer models of the human hearing system that can be used to influence processing of speech (and audio), and computer models of speech that can be used for analysis and modification of the speech signal.

By far the most important of these topics is introduced first: psychoacoustic modelling, without which MP3 and similar audio formats, and the resultant miniature music players from Creative, Apple, iRiver, and others, would not exist.

## 7.1    Psychoacoustic modelling

Remember back in Section 4.2 we claimed that this marriage of the art of psychology and the science of acoustics was important in forming a link between the purely physical domain of sound and the experience of a listener? In this section we will examine further to see why and how that happens.

It follows that a recording of a physical sound wave – which is a physical representation of the audio – contains elements which are very relevant to a listener, and elements which are not. At one extreme, some of the recorded sound may be inaudible to a listener. Equally, it is possible that a recording of sound does not contain some of the original audible features. This may be one reason why many listeners would prefer to spend a sum of money listening to live music rather than an equivalent sum to purchase a compact disc (CD) which allows them to listen to the music again and again.

Still, psychoacoustics as a study is predominantly computer-based: it rarely considers information which has not been recorded to computer, but is often used to identify parts of recorded information which are not useful to listeners.

One use of psychoacoustics is illustrated simply in Figure 7.1 showing a time domain waveform with a sudden loud pop or similar sound. A listening human ear will generally suffer from post-stimulatory masking (see Section 4.2.10) due to the sudden onset

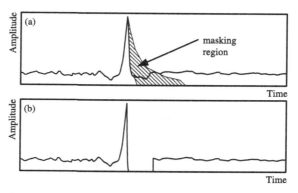

**Figure 7.1** Illustration of post-stimulatory masking in the time domain detected (a) and inaudible information stripped out (b).

of the loud sound: quieter sounds following soon after this will tend to be inaudible. Figure 7.1(a) shows a detected masking region (with sound falling in the shaded region being inaudible). Any speech processing system capable of detecting this region could practically strip out all audio within the region, as in Figure 7.1(b) resulting in something that sounds identical to the original, but is smaller in storage requirements.

For speech compression systems, this is an enormous advantage: why code/process/represent/store or transmit anything which is inaudible? The popular MP3 format detects many such inaudible regions in original music, and strips them out, compressing only the remaining audio. This mechanism allows for very low bitrate representation of music (and indeed speech): every compressed bit is fairly sure to contribute to overall sound quality.

Readers should, at this point, understand the rationale behind psychoacoustics: to obtain some advantage in coding or compression, through the consideration of the difference between what is physically stored and what is heard by a listener. We have illustrated this with a post-stimulatory masking example – but be aware that potentially most of the hearing characteristics described in Chapter 4 can lead to a psychoacoustic effect. They are most definitely not all used in practice, but at least the potential exists.

Psychoacoustics became a 'hot' topic during the 1990s, and has since been applied to almost every sphere in electronic audio systems. A comprehensive list of systems using psychoacoustic-based algorithms would be extremely long, but the following few categories encompass the vast majority:

- compression of high-fidelity audio;[1]
- compression of speech;
- audio steganography (data hiding, or 'watermarking', in audio) [1];
- active noise cancellation systems;
- speech intelligibility improvement in noise [2].

[1] Many purists maintain that high-fidelity audio *can't* be compressed, and some others argue that the long play (LP) record already represents sufficient compression! Here we take the view that such things are possible.

Despite the profusion of applications, there are relatively few fundamental psychoacoustic models, or modelling techniques used. Most of the applications employ a subset of a model predicting masking effect, an equal-loudness pre-emphasis, and an appreciation of frequency discrimination. Unfortunately, masking effect models do not often cater for temporal masking, which we presented as an example above (Figure 7.1).

By contrast, simultaneous frequency masking is well catered for in masking models. This type of masking was described and illustrated in Section 4.2.8, particularly in Figure 4.3. Simultaneous frequency is well catered for because it is relatively easily modelled by computer. Models relate to single tones, of given frequency and power, which cause nearby tones of lower power to be inaudible. In Figure 4.3, the shaded area showed the extent of the masking effect for the signal of given frequency response: in essence it is a modified threshold of audibility similar to that of the equal-loudness contours of Section 4.2.1. The difference in audibility caused by the presence of the tone is the masking effect.

Computational models exist for this masking effect due to tones. Of note is that much of the historical auditory data used to derive the computerised models were performed under well-defined and controlled conditions, with artificial signals such as white noise, and sinewaves used. While it is beyond reasonable doubt that the models describe those scenarios very well, it has not been established with the same confidence that the models accurately describe complex real sounds. In fact there is even some doubt that they can be applied to compound sounds [3]. Despite the doubts, these models are used in practice, and assume that complex sounds can be broken down into a set of tones, each of which result in a masking effect such as shown in Figure 4.3, with the overall masking effect from the sound being the summation of the separate contributions. When calculating the overall effect, it is possible to introduce nonlinear 'corrections' to the model to compensate for the fact that the effect is not a straightforward summation.

Perhaps more useful is to use a critical band model of the ear. The loudest tone in each critical band is audible, and the masking effect is the weighted sum of all sound and noise components within that band. Whilst this does not account for several auditory factors, it does model the overall situation quite well – especially for situations with a clear distinction between wanted signal and interfering noise.

In the remainder of this section we will use this approach to develop a simple but usable psychoacoustic model that has been applied, and tested, in several applications. The model involves several stages of processing. The assumption is that a particular audio signal is modelled to determine a threshold of masking (audibility) due to those sounds. Since there is very little likelihood that the absolute signal levels recorded on computer are the same as those that would impinge on the ear of a listener, then it must be stressed that – even assuming no variability among different people – the conclusions drawn by this model should be treated as being approximations. This model is provided to enable the reader to rapidly enter the field of psychoacoustics: to adjust, extend and improve on the system described. Many competing – and probably better – models exist, since this is such an active research area. Above all, the reader is encouraged to access the literature and develop their own systems.

The processing stages of the simple psychoacoustic model which will now be presented, are as follows:

1. spectral analysis;
2. critical band warping;
3. critical band function convolution;
4. equal-loudness pre-emphasis;
5. intensity-loudness conversion.

Following the model introduction we will discuss the use of the model and its applicability to speech.

### 7.1.1     Spectral analysis

Since simultaneous masking effect occurs in the frequency domain, the first step is to select a frame of audio to analyse, window it, and convert to a spectral representation. For an example frame of length 256 in MATLAB, we would do the following:

```
S=abs(fft(hamming(256).*speech));

S=S(1:128);
```

### 7.1.2     Critical band warping

Now the spectrum $S(\omega)$ needs to be warped in the frequency domain so instead of having units in hertz, it fits to a Bark scale (Section 4.3.2). This then represents the spectral index in Barks, so the effect of the critical band filters can be calculated using a Bark-domain spreading function. There are, of course, several competing models to account for spreading across critical bands from different authors. Some of the more prominent ones will be compared in Section 7.1.7, although for simplicity we shall use the approach of Hermansky here, related by the following equations [4]:

$$\Psi(\Omega) = \begin{cases} 0 & \text{for} \quad \Omega < -1.3 \\ 10^{2.5(\Omega+0.5)} & \text{for} \quad -1.3 \le \Omega \le -0.5 \\ 1 & \text{for} \quad -0.5 < \Omega < 0.5 \\ 10^{-1.0(\Omega-0.5)} & \text{for} \quad 0.5 \le \Omega \le 2.5 \\ 0 & \text{for} \quad \Omega \ge 2.5. \end{cases} \quad (7.1)$$

We had already defined two MATLAB functions to convert between Hz and Bark scales in either direction in Section 4.3.2, named f2bark() and bark2f(). We will use these to construct an analysis with 40 critical bands uniformly spaced in the Bark domain, then define critical band filters, through calculating a Bark range of each filter.

We will continue the MATLAB for this in a moment, but first consider the next two operations that must be performed in conjunction with this process.

### 7.1.3     Critical band function convolution

The critical band descriptions of Equation (4.1) need to be convolved with the warped spectrum. This generates a critical band power spectrum, $\Theta(\Omega_i)$, with the 40-band resolution mentioned previously:

$$\Theta(\Omega_i) = \sum_{\Omega=-1.3}^{2.5} P(\Omega - \Omega_i)\Psi(\Omega). \tag{7.2}$$

The scaling for an 8 kHz sampling system will be between 0 Bark (0 Hz) and 15.575 Bark (4 kHz) in equal spaced bins.

### 7.1.4     Equal-loudness pre-emphasis

The base frequency selectivity of human hearing is generally approximated by the A-weighting curve of Section 4.2.1, which itself has been nicely approximated in the frequency domain by several authors, most notably by Hermansky [4]:

$$E(\omega) = \frac{\omega^4(\omega^2 + 56.8 \times 10^6)}{(\omega^2 + 6.3 \times 10^6)^2(\omega^2 + 0.38 \times 10^9)}. \tag{7.3}$$

It should also be noted from the original source that Equation (7.3) is only accurate up to about 5 kHz (which is beyond what we require). If higher frequencies are to be used then a further term should be added to better match the upper frequency region of the A-weighting curve:

$$E(\omega) = \frac{\omega^4(\omega^2 + 56.8 \times 10^6)}{(\omega^2 + 6.3 \times 10^6)^2(\omega^2 + 0.38 \times 10^9)(\omega^6 + 9.58 \times 10^{26})}. \tag{7.4}$$

In MATLAB we can define an equal-loudness mapping (for a 128-element spectrum array):

```
for k=1:128
w=4000*2*pi*k/128;
Eql(k)=(((56.8*(10^6))+(w^2))*(w^4)) /
((((6.3*(10^6))+(w^2))^2) *
  ((0.38*(10^9))+(w^2)));
end
```

and with that we can put together the entire mapping so far into our perceptual model:

```
b4k=f2bark(4000); %Bark frequency of 4kHz
```

```
b0=f2bark(0); %Bark frequency of 0Hz
n=128; %Size of spectrum being analysed
for bi=1:40 %Assuming a Bark resolution of 40 bands
  bark=b0+bi*(b4k-b0)/40;
  wm=round(n*bark2f(bark)/(4000));
  if (wm==0)
    wm=1;
  end
  %establish limits
  w_low=bark2f(bark - 1.3)*2*pi;
  w_hig=bark2f(bark + 2.5)*2*pi;
  wl=fix(w_low/(4000*2*pi/n));
  wh=fix(w_hig/(4000*2*pi/n));
  %clip to index size
  if(wl<1)
    wl=1;
  end
  if(wh>n)
    wh=n;
  end
  %perform summation
  for wi=wl:wh
    w=wi*2*pi*4000/n;
    %Find the value of pi (from -1.3 to 2.5)
    vlu= 6*log( (w/c) + ((w/c)^2 + 1)^0.5);
    vlu=vlu-bark;
    %Look at pi & get multiplier
    mul=0;
    if(vlu<-1.3)
      mul=0;
    else
      if(vlu<=-0.5)
       mul=10^(2.5*(vlu+0.5));
      else
       if(vlu<0.5)
        mul=1;
       else
        if(vlu<=2.5)
         mul=10^(0.5-vlu);
        end
       end
      end
    end
  end
  X(bi)=X(bi)+Eql(wm)*mul*p(wi);
```

```
      end
   end
```

## 7.1.5　Intensity-loudness conversion

One final step remains, and that is to make an intensity-loudness conversion to relate the almost arbitrary units of the model to a perceived loudness scale, based on the power law of hearing. Whilst this is required for completeness, it is not needed where the model is used to compare two signals directly and the absolute difference is not required.

$$p(\Omega) = \{E(\omega)\Theta[\Omega(\omega)]\}^{0.33}. \tag{7.5}$$

## 7.1.6　Masking effect of speech

Most computational masking models have been derived for a situation where a single loud tone masks a quieter tone. Some models have been developed beyond that: for cases of noise masking tones, or even tones masking noise. Unfortunately though, evidence to suggest that the models are accurate for any and every combination of sounds is weak. Despite this the models *have* been applied, apparently successfully, to generalised audio, most notably in MP3 players.

A method of application in such generalised scenarios is shown in Figure 7.2, where a particular sound spectrum is analysed across a set of critical band regions. The sound falling within each critical band is totalled, and its masking contribution within that band determined. Then, for each band, the effect of masking spread from immediate neighbouring bands is factored in, assuming it is additive to the masking originating from the band itself. It is unusual to consider masking spread from bands beyond the immediate neighbours.

The result of such an analysis is that each critical band will have an individual masking level. Sounds within that band that are above the masking level will be audible, and sounds below the masking level will not be audible.

For application to speech, one relatively useful method of determining audibility is to look at the formant frequencies, and consider these as independent tones, most important being formants F2 and F3 which contribute more to intelligibility than the others (see Section 3.2.4). For segments of speech with no formants, the audibility of the overall spectrum (perhaps within the range 250 Hz to 2 kHz) can be determined. If the overall spectrum, or F2 and F3, are inaudible, then it is likely that the speech itself will be unintelligible. It is possible, however, that elements of the speech, whilst unintelligible, will be audible. This underlines the fact that the ability of the human brain and hearing system, to extract signal from noise, is quite awesome at times.

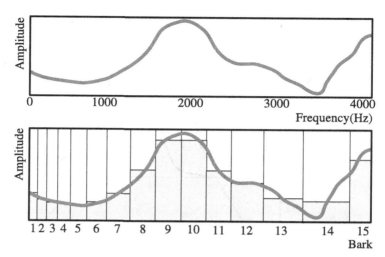

**Figure 7.2** In-band masking level derived and plotted (bottom) from an example spectrum (top) for 15 critical bands of constant Bark width.

### 7.1.7 Other critical-band spreading functions

Experimental results indicate that sound in one critical band will spread its influence into neighbouring critical bands. Given a loud sound in a particular band, depending upon how far away from the band centre we look, this will have an effect upon other nearby sounds. Many models exist to quantify this spreading for a particular sound, known variously as critical-band spreading function, noise masking curve and lateral inhibition function.

Several approximation functions were empirically derived by various authors, and are plotted in Figure 7.3. Of the curves shown, only the model of Cheng and O'Shaughnessy [5,6] attempts to account for lateral inhibition [3] where a sound in one band de-sensitises neighbouring frequencies. Jayant *et al.* [7], Virag [8] and Sen [9] report very similar curves, differing mainly in the upper frequency region. However Sen introduces corrections for both absolute frequency (the plot shows frequency relative to critical band centre frequency) and absolute power, in an attempt to improve accuracy. For the plot shown, a centre frequency of 1 kHz and absolute power of $70\,\mathrm{dB}_{SPL}$ were assumed. The former two authors do not account for absolute position in frequency or amplitude.

The curve used by Hermansky [4] has a flat-topped response that not only approximates the bandwidth of each critical band, but accounts for spreading with a dependence on absolute frequency and amplitude in a fashion similar to the model from Sen [9]. Most importantly the flat top of the function eases the computational load of the model.

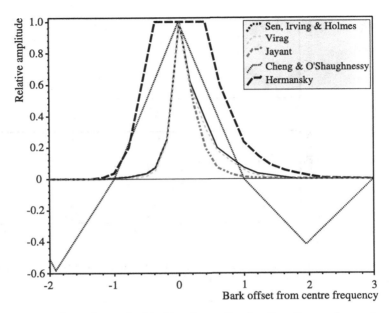

**Figure 7.3** Comparison of several critical-band spreading functions from various authors.

The models are applied on a per-band basis. The centre frequency of each critical band is used as a reference, and sound falling in that critical band will have a masking effect on other sounds as represented by the function. For example, using the Hermansky model, a sound within the critical band plotted will mask a sound 0.8 Bark lower that has amplitude 0.2 times or less.

## 7.2      Perceptual weighting

The perceptual error weighting filter (PEWF) is a common sight within speech coders. In this context it has a particular meaning which may not be quite the same as its interpretation elsewhere. Based on its context of speech coding, it makes use of linear prediction parameters, which themselves encode vocal tract information. It uses these parameters to 'strengthen' resonances, and thus to increase formant power in encoded speech.

The idea is that, since the formant regions are more relevant to human perception, the weighting process improves the perception of these. Whilst this argument is true of speech, for general music systems, perceptual weighting more often involves either the use of a perceptual model, or simply the application of a digital version of the A-weighting filter. Here we will present a typical PEWF as found within a CELP speech coder. The LPC synthesis filter is termed $H(z)$, and two bandwidth expansion factors are used, $\zeta_1$ and $\zeta_2$ with the relationship $\zeta_1 < \zeta_2 \leq 1$. Then the weighting filter $W(z)$ is defined as:

$$W(z) = \frac{1 - H(z/\zeta_1)}{1 - H(z/\zeta_2)}. \tag{7.6}$$

Remembering that the LPC synthesis filter is defined as:

$$H(z) = \sum_{k=1}^{P} a_k z^{-k} \tag{7.7}$$

then the frequency scaled version will be:

$$H(z/\zeta) = \sum_{k=1}^{P} \zeta^k a_k z^{-k}. \tag{7.8}$$

Taking Equation (7.8) in difference form, and substituting into Equation (7.6), the PEWF is quite simply realised in discrete terms as:

$$y[n] = x[n] + \sum_{k=1}^{P} a_k \{\zeta_2^k y[n-k] - \zeta_1^k x[n-k]\}. \tag{7.9}$$

This can be applied at the output of a CELP speech decoder to slightly enhance the intelligibility of voiced speech, and also be used within the CELP encoder to accentuate the importance of any formant regions within the codebook search loop (i.e. to weight the mean-squared matching process toward any formants that may be present). Typically the value of $\zeta_1$ and $\zeta_2$ is very close to unity. In the past the author has used $\zeta_1 = 0.95$ and $\zeta_2 = 1.0$ for several systems.

## 7.3    Speaker classification

Speaker classification is the automated determination of *who* is speaking. This is related to, and overlaps with, two very similar research areas of speaker verification and speaker identification. The verification task uses a-priori information to determine whether a given speaker is who he claims to be, with a true or false result. The identification task also uses a-priori information, but in this case, determines which speaker, from a set of possible speakers, is the one currently talking. The classification task by contrast is a far higher level task. It does not presuppose much, and is simply the act of placing the current speaker into one or more classes, whether a-priori information is available or not.

In practice, a-priori information would normally be available in a real system, probably captured when candidate users are registered with that system. Within such constraints, there are two further main branches to this research area: one in which the material being spoken is fixed, and the other in which the material being spoken is unrestricted. In the unrestricted case the problem is more difficult, and accuracy may well be more closely related to the amount of captured data that can be analysed than upon the accuracy of the analysis system employed.

Whichever branch is being considered, the methodology relies upon the way in which the speech of two speakers differs. If there are instances of both speakers saying the same words, either through restricting the words being spoken, or perhaps through fortuitous sample capturing, then the analysis becomes easier. Having two speakers saying the same phoneme at some time is far more likely to be achievable in the unrestricted case than speaking the same word. However this would require an accurate method of identifying and isolating different phonemes – itself a difficult task to perform automatically. In the unrestricted case, it would be possible to use information pertaining to *what* is being said as part of the analysis. At a higher language level, grammar, pronunciation and phraseology can each help to differentiate among speakers. Some of the progress in this field has been tracked by A. P. A. Broeders of Maastricht University in two papers reviewing the period 1998 to 2001 [10] and from 2001 to 2004 [11].

The restricted and unrestricted text cases mentioned above are also known as *text-dependent speaker recognition* and *text-independent speaker recognition* in the research literature. In one of the classical reviews of this field, S. Furui not only subdivides the research field similarly to the way we have discussed above, but separately discusses the ability of several processing methods [12] :

**Dynamic time warping (DTW)** where strings of features matched from speech under analysis are shifted in time, compressed and expanded to match to stored templates of those features from each candidate speaker.

**Hidden Markov model (HMM)** to match statistical property similarities between a speaker and candidates.

**Long-term statistical methods** to consider how the feature tendencies of one speaker match the candidates'.

**Vector quantisation (VQ) methods** to compile a simplified feature vector based upon the measured features from one speaker, and to match this against the stored codebook from candidates.

**Ergodic-HMM based methods** to use an approach similar to VQ, but with the statistical matching properties of the HMM to cater for patterns of temporal variation.

**Speech recognition based methods** which can, in principle, be used to detect phonemes, or phoneme classes, either for analysis of their distribution, or potentially to improve the accuracy of any of the methods already listed.

Even assuming that no background noise is present, the same microphone is used (with the same distance and orientation to the lips) and the spoken words are identical, the speech of two people may differ for many reasons including:

**Physical characteristics** such as length of vocal tract, size of nasal cavity, tooth position and so on. The effect of vocal tract length is one reason why the voices of children change, becoming less 'squeaky' as they grow older. It is most noticeable during puberty when a boy's voice may break. This is caused by his small larynx and short, thin vocal chords starting to lengthen and thicken, while simultaneously facial bones grow to create larger nasal and buccal cavities. The change does not happen overnight (although it may seem to), but shows how even relatively gradual changes in physiology can cause much more dramatic changes in speech.

**Physiological characteristics** such as the speed of tongue movement, size of mouth opening, placement of tongue, lip shaping and so on that are the features of a particular person when speaking.

**Behavioural characteristics** include the loudness and expressiveness of a person's voice. Expressiveness may well include pitch changes, and may be for linguistic reasons (accent, pronunciation, and so on). Some people speak faster than others, and would thus have a higher syllabic rate. Other people *flatten* their vowels, produce rolling /r/ sounds, guttural /g/ sounds, and so on.

However, in the computational domain, we have simply a one-dimensional signal to consider: the audio vector. The process for almost any classification task will begin similarly by performing an analysis upon the input signal to extract features. In Chapter 6 we discussed many such features. It is very likely that some of those features are more suitable for use in classifying speakers, and there are probably as many different features to choose from as there are researchers working in this field [13].

Before leaving the discussion on speaker classification, it must be remembered that there can be significant variation in the recorded voice of a person. We have already alluded to issues such as microphone response, distance to the mouth of the speaker, orientation to the speaker, and the presence of background noise. But even assuming perfectly repeatable recording, the voice may change. Some reasons for this might include:

- Long-term effects due to physiological changes such as growth, or as another example, due to smoking-induced cell damage to the throat lining.
- Long-term behavioural changes. To take one example, it has been observed by the author that those living in Japan tend to speak quietly while those living in Hong Kong tend to speak loudly, from a UK English perspective. Would I learn to speak more quietly after living in Japan for a few years, or more loudly after living in Hong Kong for a few years?
- The emotional state of the speaker, including anger, but also stress and even gentleness.
- Illness, such as influenza, sinusitis – inflamed throat, blocked nose, and so on. Inner ear infections of various forms will alter the feedback path between voice and brain that is part of the speech production regulatory mechanism, thus affecting speech.
- Time of day – again the author observes that people tend to speak more slowly in the early morning if they are unused to waking early. Waiting for the 'brain to warm up', as an explanation, may not satisfy our medical colleagues, but at least explains the effect for us.
- Fatigue or tiredness, whether mental or physical (vocal).

- Auditory accommodation – coming from a loud machine room into a quiet area often causes speakers to initially misjudge the volume of their voice. Imagine a user walking up to a microphone, extracting the in-ear headphones from their iPod and then speaking their pass phrase.
- Coffee, curry, and many other beverages or foods seem to adhere to the lining of the mouth and throat, affecting the qualities of a person's voice for some time after consumption.

To highlight the problems that these *intra-voice* changes can cause, accomplished speaker recognition researcher S. Furui admits that so far no system has succeeded in modelling these changes [12], the problem being that relaxing the accuracy requirements of a speaker recognition system to allow for variations in a user's voice will naturally tend to increase the percentage of incorrect classifications. In the vocabulary of the researchers, allowing more 'sheep' (valid users that are correctly identified) and less 'goats' (valid users that are not correctly identified), also causes more 'wolves' (invalid users that can impersonate the sheep). In addition, researchers sometimes refer to 'lambs' – the innocent sheep who are often impersonated by the big bad wolves [14].

## 7.4    Language classification

Automatic language classification, by analysis of recorded speech, has much in common with the automatic speaker classification task of Section 7.3. It can be subdivided in a similar way – namely whether there is any constraint upon what is being said and upon whether there is any constraint upon the identities and number of persons speaking. More importantly, the base set of analysis features and techniques is common.

In an extensive review of the language identification research field, Zissman and Berkling [15] cite four auditory cues that can distinguish between languages:

**Phonology** (see Section 3.2) generally differs in that not all languages comprise the same set of phonemes, and undoubtedly they are used in different sequences and arrangements between languages.

**Morphology** meaning that languages tend to have different, but often similar, lexicons. By and large, languages derived from the same root will share a more common morphology. However imported or shared words blur this distinction.

**Syntax** differs in style, sequence and choice of framing words. For example, some languages tend to prefix nouns with prepositions, others do not. Some languages, such as Malay, have far more word repetitions than others, such as English.

**Prosody** is the rate, spacing and duration of language features.

Although the researchers do not normally discuss the issue, the research field is confronted by a rather difficult problem: those brought up speaking one language may retain the prosody and even syntax when speaking another language. Anyone who has travelled to Pakistan or India and heard locals speaking English would be struck how the prosody of the language had become localised. Similarly, syntactic differences lead many Mandarin Chinese speakers who are starting to learn English to ignore rules of plurality and tense. Greeks speaking English may well include too many definite articles 'having the knowledge is the good thing'. Conversely the author has been accused at times of speaking both Chinese and Greek like an Englishman.

The difficulty of speakers bringing the prosody from their native tongue to another language is analogous in many ways to the issue of one speaker impersonating another to a speech classification system. It is a source of inaccuracy, but is not necessarily a major problem. Imagine a language recognition system built into a call centre that routes incoming callers to customer service operators able to speak their language. That a Greek speaking English with Greek syntax would be misinterpreted as speaking Greek is not a major problem. However, if he were misinterpreted as being a Japanese speaker, that may well be a major problem.

Continuing on to the techniques used for the classification, Zissman and Berkling [15] provide an overview of the various techniques, and also go on to provide a comparative evaluation of each of these. Their techniques are:

**Spectral-similarity** approaches based upon the different spectral characteristics of languages, or upon the statistics of the spectral changes during speech for each language.

**Prosody-based** approaches which look at the timings and duration of captured linguistic features.

**Phone-recognition** approaches, similar to the speech recognition approach in the speaker classification task, there are differences in the relative frequency, time-distribution and sequencing of phone features.

**Multilingual speech units** called *poly-phones*, *mono-phonemes* or *key-phones* are the widest possible set of phonemes represented by all languages under consideration (poly-phones), or just the most important ones in terms of relative frequency or importance to classification (mono-phonemes/key-phones). These can then be detected by an automatic speech recognition system, with the output analysed statistically.

**Word level** approaches can search for the occurrence of key words. These systems can even be trained, without having prior knowledge, to identify and select the key words.

**Continuous speech recognition** where, at the extreme, parallel speech recognisers for each of the candidate languages compete to provide the best (most sensible) transcription. In many ways this is a brute-force approach, that would only be sensible for a relatively small set of candidate languages.

As alluded to above, the driving forces behind automated language classification are predominantly telephone call centre-related at the present time. The advent of a Star Trek universal translator (or, if you prefer, the in-ear Babelfish) is still some years away, and the present bias of the Internet is toward visual media rather than auditory communications. Having said that, for single-digit sized groups of languages, classification accuracy can exceed 90%, indicating that the technology is maturing quickly.

While neither automatic speaker recognition nor automatic speech recognition can approach the abilities of the average human, automatic language recognition systems exceed average human abilities (partly because few humans can speak 10 languages). In this way, despite the disadvantages of the technology developed to date, it can find ready application outside the research laboratory.

## 7.5    Speech recognition

Entire texts have been written on the subject of speech recognition, and this topic alone probably accounts for more than half of the research and development effort in the fields of speech and audio processing. The ability to communicate to a computer, without the use of a keyboard, has been around for almost as long as electronic computers have been with us. Yet, despite the significant research effort, speech recognition systems are elusive apart from in a few niche areas.

### 7.5.1    Types of speech recogniser

Speech recognition covers a wide field, and is normally categorised by a few key descriptive phrases:

**Automatic speech recognition (ASR)** describes a system that can recognise speech without additional user input.

**Continuous speech recognition** describes a speech recognition system that can recognise continuous sentences of speech. In theory this would not require a user to pause when speaking, and would include dictation and transcription systems. The alternative is a **discrete word** recognition system, used primarily for handling vocal commands, that recognises single words delimited by pauses.

**Natural language processing (NLP)**, whilst not strictly limited to speech, describes the computational methods needed for a computer to understand the meaning of what is being said, rather than simply knowing what words have been said. For an automated transcription system, the meaning may be irrelevant, but to create a virtual butler able to cater to human needs, the meaning of what is said would be important.

In general, we will consider primarily the case of discrete word recognisers, since this is a lower-level recognition task, closer to the physical parameters being analysed.

### 7.5.2    Speech recognition performance

Established researcher Victor Zue and colleagues have identified several parameters that can be used to characterise speech recognition systems and their performance (given in Section 1.2 of a wide-ranging survey report [16]). Based upon this work, Table 7.1 lists several characteristic parameters of these systems.

**Table 7.1.** Speech recognition system parameters.

| Parameter | Typical range |
|-----------|---------------|
| Speech type | Single words–continuous sentences |
| Training | In advance–continuous |
| Users | Single–open access |
| Vocabulary | Small–large |
| SNR | Low–high |
| Transducer | Restricted–unrestricted |

Many, if not all, recognition systems require some form of training which acclimatises the system, either to the speech of a particular individual, or to a group of individuals. This can be accomplished in advance where the speaker (or speakers) are known, or otherwise could be an 'on-line' gradual training during operation. The question of whether the system has been designed to operate for single users, small groups, or for unrestricted users evidently impacts the training methods employed. Of each of these parameters so far, a single word, pre-trained, single user system, is by far the simplest system to design with reasonable accuracy. Any deviation from this simple combination will incur a penalty in terms of either design complexity, reduced accuracy, or both. Recognisers that use rules of syntax, perhaps some sort of artificial grammar, might well benefit from being supplied with continuous speech, but by and large, single words are easier to recognise.

In terms of vocabulary size, it is reasonable to assume that the larger the vocabulary, the more difficulty any system will have in accurately detecting a word. In fact, the same is true for human speech. Figure 7.4 shows a plot resulting from a re-examination of the data shown in Figure 3.6 (Section 3.3.4), which itself is derived from experimental results obtained over half a century ago by Miller *et al.* [17]. This figure plots results for various signal-to-noise ratios of speech, as heard by human listeners, and excludes the extreme accuracy points above 85% or below 20% when other effects such as saturation come into play. The 'recognition accuracy' or articulation index is plotted against the logarithm of the vocabulary size, showing a clearly logarithmic relationship.

In fact, the logarithmic relationship has been shown to be present in some of the largest variable-vocabulary recognisers, such as the famous Bellcore telephone directory assistance system [18].

It should be noted that the continuous speech recognition research field also includes topic-specific recognisers, which are trained with a vocabulary of subject-specific words (such as a recognition system aimed at understanding medical terminology, or one trained on legal terminology). Another subset in the research field is systems which can attempt to determine the topic of speech under analysis, and switch to the appropriate vocabulary database as required.

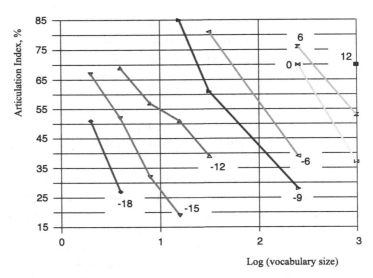

**Figure 7.4** Plot of articulation index (a measure of recognition accuracy) versus the logarithm of vocabulary size for speech recognised at various levels of additive white noise, with SNR ranging from −18 dB to +12 dB.

The process of actually capturing speech input for a recogniser is extremely important, affecting not only the signal-to-noise ratio, but also the characteristics of the sound. Telephone-based systems must cater for the low bandwidth and low quality telephone signal, whereas microphone based systems depend for their performance in part upon the distance between the microphone and the mouth of the speaker.

It is likely that any user of commercial ASR systems would have faced the problems of background noise, microphone placement, gain, and so on. Having trained a system to a particular voice so that accuracy levels of over 90% can be achieved, an issue as simple as changing microphone placement, or the addition of even quite low levels of background noise, could reduce accuracy by as much as 20%. The presence of music, speech or loud background noises may well degrade performance far more.

The presence or absence of background noise is a critical operational factor, and recognition systems designed to work with headset-mounted microphones or similar will naturally perform better than those capturing speech from a transducer located far away from the mouth of a speaker. In the latter case, directional microphones can 'steer' themselves to some extent in order to avoid background sounds. Other recognition systems have made use of non-speech cues to improve recognition performance, including: video images of a speaking mouth, ultrasonic echoes of the mouth, body gestures, facial expressions, and even nerve impulses in the neck. Clearly the speech recognition research field is both very active and very diverse.

## 7.5.3    Practical speech recognition

Practical ASR systems have some generic structure, although the details will vary quite widely. A block diagram of a generic system, shown in Figure 7.5, shows input speech is first cleaned up by a pre-processing system before a feature vector is extracted. The pre-processing may take the form of filtering, probably windowing and normalisation, and some method of segmentation.

Following pre-processing, features are extracted from the speech. There are many possible features which can be used, including LPCs, LSPs, cepstral coefficients, spectral coefficients, and so on, although Mel-Frequency Cepstral Coefficients (MFCC) are probably the most popular at present, and there is of course no reason why the vector needs to contain just one feature. Each feature may include several tens of coefficients, and be updated every 20 ms.

In the simplest of systems, these features can then be compared, in turn, to a large set of stored features (an acoustic model). A distance measure (perhaps the Euclidean distance, but more often a weighted distance measure and very commonly these days, the role is taken over by a hidden Markov model) is computed for each of the stored features, and a probability assigned to each one. This probability identifies how well the current speech segment matches the stored features, and naturally the highest probability match is the best one.

However there is another level of refinement possible beyond this one, and that is to apply a language model (also shown in Figure 7.5), to weigh the probabilities of the top few matches from the acoustic comparison based upon their adherence to language rules. For example, if the highest matching feature vector is found to be something disallowed in the language being spoken, then it probably should be rejected in favour of the second highest matching feature vector.

A dictionary can be used to refine the matching further: only phonetic combinations found in the dictionary are allowed.

Evidently, with the possibility of several hundred stored feature vector templates in the acoustic model, a similar number in the language model, and perhaps more in the dictionary, this whole matching process can be very slow. This is one reason why the vocabulary should be restricted, but also why the size of the feature vector should be minimised where possible. Much research has been done on restricting the amount of searching that may be necessary during the matching process.

The language model, as described, considers the probability that the current speech is correctly matched given knowledge of the previous unit of matched speech. In general this history can extend back further than to just the previous sound. An *n-gram language model* looks back at the past *n* speech units, and uses these to compute the probability of the next unit out of a pre-selected set of a few best matches from the acoustic model. Of course, this again increases computational complexity, but significantly improves performance (especially in more regular languages such as Mandarin Chinese). The units under consideration in the *n*-gram language model could be phonemes, words, or similar, depending upon the application, vocabulary size, and so on. In a non-regular language

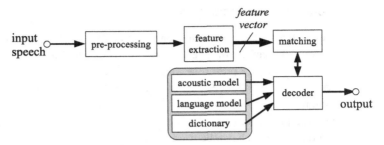

**Figure 7.5** Block diagram of a generic speech recognition system, showing input speech cleaned up and filtered in a pre-processing block, feature extraction, and then the matching and decoding processes driven from predefined models of the sounds, language and words being recognised.

such as English, the number of phonemes in different words can vary significantly, so again a dictionary can be used to adjust the parameter $n$ in the $n$-gram language model.

The output of the ASR system could be given as a string of phonemes, but is more usefully delivered as a sequence of recognised words, and again this depends upon the particular application and configuration of the system. The models of speech themselves, namely the acoustic model, language model, and dictionary as shown in Figure 7.5, are particularly important in an ASR system. A sound missing from the acoustic model, language features not covered by the language model, and words not in the dictionary cannot be recognised. Although the dictionary is often created from predefined word lists, the two models are usually the result of training. Whilst it is theoretically possible to define language and acoustic rules by hand, it is far easier and more accurate to train a system using representative speech to build up these models statistically. For a system operating with different speakers, it can be better to detect who is speaking (see Section 7.3) and then switch to an individual acoustic model, than it is to have one big acoustic model to cover everyone. Similarly, for systems encompassing several different languages or dialects (see Section 7.4) it can be better to detect these and switch language models appropriately.

Up to this point, we have discussed ASR systems in general. However it is instructive to briefly turn our attention to a particular example of state-of-the-art speech recognition: Sphinx.

The open-source Sphinx recogniser, originally developed at Carnegie Mellon University in the USA, is one of the best examples of a flexible modern speech recognition system. It can be used for single word recognition, or expanded up to large vocabularies of tens of thousands of words, can run on a tiny embedded system (PocketSphinx) or on a large and powerful server (which could run the Java language Sphinx-4), and is constantly updated and evaluated within the speech recognition research field.

Sphinx-4, in common with most current ASR implementations, relies upon hidden Markov models to match speech features to stored patterns. It is highly configurable, and incredibly flexible – the actual feature used can be selected as required. However the one that is most commonly extracted and used for pattern matching purposes is the Mel-Frequency Cepstral Coefficient (MFCC) [19].

This flexibility extends to the pre-processing sections, the 'FrontEnd' where a selection of several different filters and operations can be performed singly or chained together, and also to the so-called 'Linguist' which is a configurable module containing a language model, acoustic model and dictionary. The linguist is responsible for consulting these based upon a particular feature vector, and determining which subset of stored patterns are compared with a particular feature vector under analysis.

Sphinx-4 has been tested extensively using industry-standard databases of recorded speech, which are commonly used by ASR researchers to compare the performance of systems. Accuracy rates of over 98% are possible for very small vocabularies (with a response time of 20 ms), over 97% for a 1000 word vocabulary (in 400 ms), and approximately 81% for a 64 000-word vocabulary (below 4 s) [19]. These figures are assumed to be for high SNR cases.

## 7.5.4    Some basic difficulties

Although we have looked at the main parameters related to speech recognition, there are several issues that speech recognition systems in general need to cope with. These may include:

**Voice activity detection (VAD)**, also known as a voice operated switch (VOS) is a device able to detect the presence of speech. It would serve no purpose for an ASR system to attempt the computationally intensive task of trying to recognise what is being said when no speech is present, and thus the ability to accurately detect speech is required. However this is not a trivial task, and is in fact a research area in its own right.

**Segmentation** of speech into smaller units is often required in processing systems. Whilst this is generally based on fixed size analysis frames when performing general audio processing (see Section 2.4), in ASR systems, segmentation into words, or even into phonemes, may be required. Again, this is non-trivial, and is not simply a matter of searching for gaps within continuous speech, since the gaps between words or sentences may on occasion be shorter than the gaps within words.

**Word stress** can be very important in determining the meaning of a sentence, and although it is not captured in the written word, is widely used during vocal communications. As an example, note the written sentence 'He said he did not eat this' and consider the variations in meaning represented by stressing different words:

**He** *said he did not eat this* : indicating that someone else said so
*He* **said** *he did not eat this* : indicating that you probably do not believe him
*He said* **he** *did not eat this* : indicating that someone else did
*He said he* **did** *not eat this* : indicating that he is, or will be eating this

He said he did **not** *eat this* : indicating an emphatic negative

*He said he did not* **eat** *this* : indicating that he did something else with it

*He said he did not eat* **this** : indicating that he did eat something, but not this

and of course several of the above stresses could be used in combination.

**Context** was discussed in Section 3.3.4, and although it cannot be relied upon in all cases, it can be used to strengthen recognition accuracy upon occasion.

## 7.6        Speech synthesis

In the wider sense, speech synthesis is the process of creating artificial speech, whether by mechanical, electrical or other means. There is a long history of engineers who attempted this task, including the famous Austrian Wolfgang von Kempelen who published a mechanical speech synthesiser in 1791 (although it should be noted that he also invented 'The Turk', a mechanical chess playing machine which astounded the public and scientists across Europe alike for many years before it was revealed that a person, curled up inside, operated the mechanism). However, the more sober Charles Wheatstone built a synthesiser based on this work of von Kempelen in 1857, proving that this device at least was not a hoax.

These early machines used mechanical arrangements of tubes and levers to recreate a model of the human vocal tract, generally fed through air bellows. Different combinations of lever settings could cause the systems to create vowels and consonants on demand. All required an intelligent human to learn the operation of the system in order to decide which sounds to sequence together to form speech. In fact, much the same methodology of stringing together different phonemes is still in use today, having survived the transition to electrical systems in the 1930s through electronic systems in the 1960s and into computers. Of course, speech synthesis researchers would argue that both quality and usability have improved significantly over the past 300 years.

Today, there are three broad classifications of speech synthesisers, namely text-to-speech systems (TTS), phonetic or linguistic transcription systems, and simple playback systems. We will briefly overview each of these in turn.

### 7.6.1        Voice playback systems

Simple playback systems, often used for telephone voicemail menus, record words or sentences for later playback. An example would be systems which record spoken digits and then replay them in different sequence to generate larger numbers. Although the quality of the speech itself can be high, these systems do not sound particularly natural because the intonation of various words does not always match listener expectations. In addition these systems are not particularly flexible: they cannot create new words, and require non-volatile storage of every word in their vocabulary.

Advances in speech quantisation, in particular the CELP analysis-by-synthesis systems of Chapter 5, allow stored audio to be compressed in size, such that storage requirements, even for quite a large vocabulary, are not excessive. Otherwise, basic LPC parameters can be stored along with a pitch parameter to enable recreation of speech. Furthermore, advances in the processing of speech can allow for post-processing of a stitched-together sentence to improve naturalness – for example by imposing an overall pitch contour (with an end-of-sentence downward tail – or upward for a fake Australian accent).

At a basic technology level, stored voice playback is an example of a *concatenative* system which concatenates, or strings together sequences of sounds to synthesise an output [20]. Word-level or sentence-level concatenation is a far simpler, but less general speech synthesis solution than phoneme concatenation, which we will discuss next.

## 7.6.2    Text-to-speech systems

Text-to-speech describes the process of turning written words into audible speech. A readable overview of these systems is provided by Dutoit [21]. At their simplest, these systems can operate with only single words, and although they can use voice storage and playback, most normally use either stored or generated phonemes. More complicated systems handle entire sentences at a time.

What is needed at a word level then is a dictionary or heuristic which relates each written word to a sequence of phonemes, in effect a rule to pronounce each word. In English, this is a non-trivial task, because spelling is often not phonetic: there are many words which must be pronounced in a way contrary to a phonetic reading of their spelling. In Chinese, the task is far easier on the one hand because, apart from a few characters having dual pronunciation (normally distinguishable through context), there is a straightforward mapping between the character to be read and a pronunciation. However the exact pronunciation must be stored somewhere for whatever characters are supported – which would be at least 3000 for a basic newspaper, and rising to 13 000 for scholarly works.

Moving back to English, it is common in TTS systems, including most commercial speech synthesisers, for there to be procedures which guess at a phonetic spelling, with a dictionary to override this guess in irregular cases. In fact early-years school children tend to learn their English in a similar way: if in doubt pronounce phonetically, but learn any exceptions to the rule. The primary difference being that TTS systems do not learn new pronunciations when listeners laugh at their incorrect attempts.

## 7.6.3    Linguistic transcription systems

Text-to-speech systems are all very well, but humans do not simply read words in isolation – and that is one reason why the word playback systems of Section 7.6.1 tend to sound unnatural. Humans modulate their speech over a sentence, based upon the syntax of what is being said. Most speakers also add stress and intonation differences to particular words, either to change the meaning of what is being said (see the sentences reproduced with different stressed words in Section 7.5.4), or at least to convey emotional information

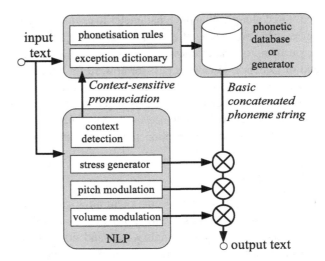

**Figure 7.6** Block diagram of a generic speech synthesis system, showing input text being analysed phonetically to output concatenated strings of phonemes (top) and linguistically through a natural language processor (bottom left) to firstly adjust for any context-sensitive pronunciation changes, and secondly to add stress, intonation, pitch and volume changes to the phoneme string to account for sentence structure, punctuation and any words that may need to be emphasised.

such as interest, distaste, happiness, and so on. Modulation could be frequency domain (pitch), time domain (rate of speaking, pauses between words, and so on) or in more complex ways that change the perceived sounds.

Anyone who, like the author, has endured listening to overlong monotonous monologues in academic conferences, should value the presence of such intonation changes and modulations. As a practical demonstration, read the sentence 'Friends, Romans, countrymen, lend me your ears' aloud. Next re-read the sentence with uniform spacing between words, and in a monologue. Most would agree that this changes the impact of the sentence somewhat.

Having demonstrated the importance of stress, intonation, pitch and pace in speech, we now need to acknowledge that these aspects are subjective elements added by a speaker, and which are not represented in the basic text. Thus even a TTS system that can reproduce single words so good they are indistinguishable from a human speaker's, would fail to produce a natural output when the words are joined together, unless the issues mentioned are solved.

For natural sounding speech synthesis, we can say that in general, as many of these elements should be incorporated as is possible. Two methods exist for incorporating these elements, firstly the use of a transcription system that includes the required information (to replace bare text input), and secondly a method of extracting such information from the text automatically.

The first system is not truly TTS since the input is not just text, but text plus linguistic markers. Stress markers can be found as accents within the international phonetic alphabet, and would normally be added by an expert transcriber (although some experimental systems exist which attempt to add this information automatically, and perhaps future speech recognition systems would be capable of recognising this information?). For a more complete system, pitch variations, prosodic information, volume modulation and so on would also need to be added. For the foreseeable future, this task probably requires human intervention, and thus systems that require linguistic marker information are better suited to small constrained vocabulary systems.

The second method of improving naturalness requires a machine to 'read' a sentence without a-priori knowledge of stress and intonation patterns, to extract such information from the text, and to add this into its reading. This is much the same task that a human speaker faces when given a paragraph to read out loud: to scan the text, and decide which words to stress, where to speed up, increase volume, pause and so on. This task is no longer considered to be within the domain of speech synthesis since it is one of interpretation, it is part of *natural language processing* (NLP) research.

Figure 7.6 provides a block diagram of a generic concatenative speech synthesiser of the second type mentioned here. A database of stored phonetic sounds is strung together by the system to match the phonemes of the input text. For some words, the phonetic sequence is generated through phonetisation rules, but can be overridden by a dictionary of exceptions to the rule. Occasionally, context would dictate different pronunciation of a word, and this would be detected through the use of an NLP system. NLP is also used to identify places in the recreated phoneme string to add in stress patterns, probably including changes in prosody, pitch variation, volume variation, and so on. It should of course be noted that this is just one possible method of performing synthesis: there are countless alternatives represented in research literature [21].

### 7.6.4 Practical speech synthesis

The Festival speech synthesis system by the University of Edinburgh Centre for Speech Technology Research is probably the most common synthesis system in use for research today. It is free software, distributed open-source style, which researchers and developers can download, test and modify as required [22]. It offers a highly configurable 'general framework' for speech synthesis research, as well as a fully working system that can synthesise speech in English, Welsh, American English and Spanish.

Festival is capable of uttering whole sentences, which it can do by formulating a grammar or syntax structure. In fact, at the word level, a sequence of words in time is related, inside Festival, to other attributes of those words. Words themselves comprise phones. Word functions are identified, and used to allow different rules to apply to those words. In this way, Festival could for example utter the subject of a sentence at a slower speed, and a verb slightly louder than other words, since basic speaking information can be calculated, in real time, from syntactic information [23].

In common with many systems, Festival uses a pronunciation lexicon. This looks up pronunciation of a word (or a word part, or a near match). Unknown words are pronounced by a letter-to-sound rule. A significant amount of work has also been put into the reading of various punctuation marks and abbreviations in Festival, which improves the perceived naturalness when reading real-world text, such as emails, considerably. For further improvement, Festival attempts to speak syllables rhythmically, adjust speaking duration, modify pitch to fit accents and any tonality which may be required. Users may also inform the synthesiser concerning the class of what is being spoken – such as words, phrases, raw text, and so on, and provide relevant information to specify phrasing, timings and even tone changes [22].

## 7.7     Stereo encoding

Stereo means something that is stored or reproduced in more than one channel. It can also be referred to as *stereophonic*. This was contrasted to the technical term 'binaural' in Section 4.2.13 which means something pertaining to both ears. In audio terms, stereo normally refers to two channels of audio, one for a right loudspeaker and one for a left loudspeaker, but many other multichannel arrangements exist.

The basic idea of stereo audio relates to the fact that most of us have two ears which we use to differentiate between sounds, particularly regarding source placement. In nature, a sound to the left of our heads will be heard by the left ear slightly before it is heard by the right ear, and slightly louder. This tiny delay, or phase difference, between the two versions of sound heard by the ears is sufficient to allow the brain to calculate where the sound originated from. We are not conscious of this calculation since it is performed automatically, but we can demonstrate the effect very easily using MATLAB.

First we will create a sound made from two related sinewaves. These need to be sampled at a higher frequency than some of our demonstrations because of the more fine control over timing we can achieve. In this case we will use a 44.1 kHz sample rate to create a 440 Hz note in the usual way:

```
Fs=44100;
Ft=440;
note=tonegen(Ft, Fs, 0.5)
```

Next we will replay two versions of this using soundsc. In MATLAB, stereo sound is represented by a two-column matrix of samples. The first column holds samples for the left channel, and the second column holds samples for the right channel. So we will create two two-column matrices of audio, one in which the left channel is delayed slightly, and one in which the right channel is delayed slightly:

```
s1=[[zeros(1,20), note];[note, zeros(1,20)]];
s2=fliplr(s1);
```

where fliplr() is used to switch the left and right channels around. Next we will listen to those two sounds. You will definitely need a computer capable of replaying in stereo (and most are these days), and quite likely will require a pair of stereo headphones to really appreciate the stereo placement:

```
soundsc(s1, Fs);
soundsc(s2, Fs);
```

When listening, the first of these should sound like it is coming from a source located to the right, and the second should sound like it is from a source located to the left.

Also mentioned in Section 4.2.13, if headphones are not used, then both ears will hear a mixture of the two sounds. This is a slightly more complicated proposition, but again, good quality loudspeakers can accurately simulate the sound localisation. If in doubt, try it.

### 7.7.1   Stereo and noise

Interestingly, it is not necessary for the two sounds to be related for the brain to hear in stereo. In this example, we will create two vectors of random noise. First we will play the same vector to both ears:

```
r1=0.2*rand(1, Fs*0.5);
r2=0.2*rand(1, Fs*0.5);
%Note we use sound rather than soundsc to save our ears
sound([r1;r1], Fs);
```

This should sound as if the noise is located in between the ears. Next, we will play different random noise signals to the two ears:

```
sound([r1;r2], Fs);
```

Having uncorrelated signals in each ear causes the brain to assume that the signals are unrelated, and separated. The sound now appears to be spatial in nature – stereo noise in effect. As an aside, this is one reason why some audio purists still prefer to listen to music from a turntable: the noise from a stereo vinyl record is predominantly common to both channels, and thus is not stereo. The noise may appear to hover in between the loudspeakers, leaving the listener free to enjoy the spatially separated sounds.

### 7.7.2   Stereo placement

An arbitrary spatial placement for stereo sounds can be calculated from a physical representation of the system we are attempting to model. Given a sound some distance $d$ away from a listener, and at an angle $\theta$ away from the location directly ahead, sounds will need to travel further to reach one ear than to reach the other. This is shown in Figure 7.7 where the left ear will receive sound from the point identified slightly later than the right ear will.

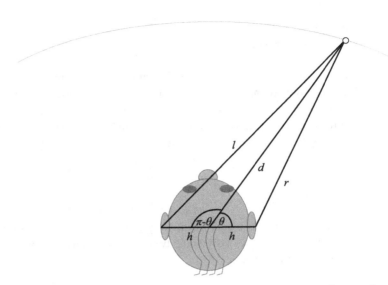

**Figure 7.7** Diagram of angle-of-arrival of sound from a single source at distance $d$ and angle $\theta$ causing a path distance difference for sound reaching the two ears.

If we also denote the head radius as $h$ (thus the distance between the two ears as $2h$), then we can use the cosine rule to calculate exactly the path difference travelled by audio reaching each ear.

$$l^2 = d^2 + h^2 - 2dh \times \cos(\pi/2 + \theta) \tag{7.10}$$

$$r^2 = d^2 + h^2 - 2dh \times \cos(\pi/2 - \theta). \tag{7.11}$$

Then assuming location at sea level where the speed of sound is approximately 350 m/s, sounds will reach the left ear in $l/350$ s, and the right ear in $r/350$ s. At a sample rate of $Fs$ samples per second, the difference in number of samples between the two would be $Fs \times 350/(l - r)$. In MATLAB we could then replay a sound, in stereo, with one of the channels delayed by this many samples to cause the brain to interpret the sound as coming from angle $\theta$. We use precisely this calculation in the following MATLAB code to simulate a sound moving around the head, additionally with a rudimentary scaling system whereby the amplitude heard by each ear is inversely proportional to the distance to the simulated sound source (so that the ear closer to the sound hears a slightly higher amplitude signal than the one further away):

```
d=5;    %distance
h=0.1; %head radius
Fs=44100;
Ft=600;
note=tonegen(Ft, Fs, 0.10);
note=note+tonegen(Ft*2, Fs, 0.1);
```

```
%Speed of sound
Vs=350; %m/s
ln=length(note);
%Cosine rule constants
b2c2=d^2 + h^2;
b2c=2*d*h;
for theta=-pi:pi/20:pi
   %Calculate path differences
   lp= b2c2+b2c*cos((pi/2)+theta);
   rp= b2c2+b2c*cos((pi/2)-theta);
   %Calculate sound travel times
   lt= lp/Vs;
   rt= rp/Vs;
   %How many samples is this at sample rate Fs
   ls= round(Fs*lt);
   rs= round(Fs*rt);
   %Handle each side separately
   if(rs>ls)   %right is further
     df=rs-ls;
     left=[note, zeros(1,df)]/ls;
     right=[zeros(1,df),note]/rs;
   else          %left is further
     df=ls-rs;
     left=[zeros(1,df),note]/ls;
     right=[note, zeros(1,df)]/rs;
   end
   %Create the output matrix
   audio=[left;right];
   soundsc(audio, Fs);
   pause(0.1);
end
```

### 7.7.3    Stereo encoding

Stereo is normally stored simply as either interleaved left and right samples, or as a two-column matrix as in MATLAB, but in professional audio fields, there are far more possibilities. Infobox 7.1 describes some of these.

Within audio systems, the traditional approach is to maintain left and right channels as separate streams of data. Gain, filtering, and so on, would be performed independently on each. However, systems which purport to enhance stereo separation, surround sound systems and spatial audio systems would each mix together the stereo channels with different, possibly dynamic, phase relationships. In compression systems, it is also common to encode stereo channels jointly. In many cases this is simply to encode middle and side channels, created from the difference between the left and right channels, separately (see the Infobox Common stereo audio standards), under Mid-side stereo, where the mid channel is $(R - L)/2$ and the side channel is $(L - R)/2$, performed to improve quantisation performance. On the other hand, some systems perform a type of joint stereo encoding in which the lower frequency information is stripped from both the left and right channels (leaving them simpler, and thus able to be compressed using fewer bits), and then encoded separately.

Although the term 'joint stereo encoding' is frequently misused, especially in the speech compression community related to MP3 encoding, the idea of joint stereo encoding is that humans are not really able to localise low frequency sounds, and hence there is no need to represent them in stereo.

A very simple experiment to show this is to repeat the MATLAB demonstration of the previous section with much lower frequency sounds, perhaps 40 Hz or so (if your computer can reproduce such a low frequency). Also, the prevalence of single sub-woofer setups in modern audio systems is due to the fact that two would be redundant: a house-shaking bass sound permeates an entire room and the brain cannot easily localise the source.

---

**Infobox 7.1** Common stereo audio standards

There is a plethora of stereo audio standards, and non-standards in use around the world today. While the vast majority of stereo music is recorded in a straightforward R + L arrangement, including CDs, MP3, and so on, digital audio coding used on DVD and more advanced media, tends to be more complex. Still more professional systems, as used in cinemas, add to the complexity. Classified under the general term 'surround sound', these advanced systems include the following notable arrangements (any trademarks and trade names mentioned are the property of their respective owners):

| Name | Front | Middle | Rear | Notes |
|---|---|---|---|---|
| Mono | C | – | – | |
| Stereo | L, R | – | – | |
| Mid-side stereo | $(R - L)/2$, $(L - R)/2$ | – | – | |
| Dolby surround | L, R | – | S | |
| Quadraphonic | L, R | – | L, R | |
| Dolby Pro Logic | L, R, C | – | S | uses five speakers |
| Dolby Pro Logic II | L, R, C | LFE | L, R | known as 5.1 |
| Dolby Digital, DTS | L, R, C | LFE | L, R | known as 5.1 |
| Dolby Pro Logic IIx | L, R, C | L, R | C | plus LFE |
| Dolby Digital Plus | L, R, C | L, R, LFE | L, R | known as 7.1 |
| 10.2 channel | $2 \times$ L, C, $2 \times$ R | $5 \times$ S | $2 \times$ LFE | two height channels |
| 22.2 channel | arranged in 3 layers: 9 top, 10 middle, 3 lower, 2 woofers | | | |

**L** = left channel, **R** = right channel, **S** = surround channel, **C** = centre channel
**LFE** = low frequency effects channel

There are many more specialised systems in use around the world. The urge to add more and more channels appears to be unstoppable, despite the fact that humans have only two ears with which to hear sounds.

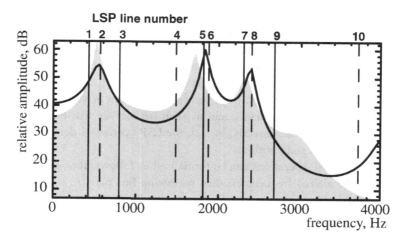

**Figure 7.8** A spectrum (in grey) overlaid with an altered set of LSPs and the spectrum resulting from the alteration. The original spectrum was given in Figure 5.10 on page 107. Alterations include line widening and shifting (see text).

## 7.8 Formant strengthening and steering

As we know from previous chapters, LSPs are convertible to and from LPC coefficients, and when in use in speech compression systems, LSPs will be quantised prior to transmission or storage. It is thus common to find that the decoded LPC coefficients differ from the original LPC coefficients in some ways. The use of an intermediate LSP representation for quantisation ensures that instabilities do not occur.

A comparison of the original and final spectra shows differences in the immediate frequency regions of the lines that were changed most. From these observations it has been found possible to alter the values of particular LSPs to change the underlying spectral information which they represent.

To illustrate this, Figure 7.8 plots the original spectrum from Figure 5.10, overlaid with a spectrum derived from an altered set of LSPs. The changes in the new spectrum came about through increasing the separation of LSP pair {1:2}, decreasing the separation of line pair {5:6} and shifting line 10. The increased separation of pair {1:2} resulted in a wider, lower amplitude spectral peak between them, whereas the decreased separation of pair {5:6} plus a slight upward frequency translation, caused a sharper peak between them, now at a higher frequency. Finally, moving line 10 closer to the Nyquist frequency of 4 kHz, caused a spectral peak to form at that frequency.

It was probably Paliwal [24] who first reported that the effects on the underlying spectrum of modifying a line, are predominantly confined to the immediate frequency region of that line. However this observation is correct only in so far as obvious changes are concerned, since amplitude changes in one region will always cause compensatory power redistribution in other regions.

Despite this, as long as line alterations are minimal, the effects on other spectral regions will be negligible. This is saying that, for small movements, and small movements only, localised spectral adjustments can be made through careful LSP movement.

The example of Figure 7.8 showed a spectrum of voiced speech. The three spectral peaks represent formants, and as such we can see that the operations we performed affected formants directly. In fact, the LSP operations demonstrably adjusted formant bandwidths and positions.

The LSP operations can be formalised as follows. If $\omega_i$ are the LSP frequencies and $\omega'_i$ the altered frequencies, then narrowing line pair $\{i : i+1\}$ by degree $\alpha$ would be achieved by:

$$\omega'_i = \omega_i + \alpha(\omega_{i+1} - \omega_i) \tag{7.12}$$

$$\omega'_{i+1} = \omega_{i+1} - \alpha(\omega_{i+1} - \omega_i) \tag{7.13}$$

and increasing the frequency of line $k$ by degree $\gamma$ may be achieved with:

$$\omega'_k = \omega_k + \omega_k(\gamma - 1)(\pi - \omega_k)/\pi. \tag{7.14}$$

When altering the line positions it is important to avoid forming unintentional resonances by narrowing the gaps between lines that were previously separated. This problem may be obviated either by moving the entire set of LSPs, or providing some checks to the adjustment process [25]. In the former case, movement of lines 1 and 10 closer to angular frequencies of 0 and $\pi$ may also induce an unintentional resonance. Equation (7.14), designed for upward shifting, progressively limits the degree of formant shift as a frequency of $\pi$ is neared. A similar method may be applied to downward shifting.

In MATLAB, a simple program to detect the three narrowest pairs of lines in a set of LSPs (in vector lsp), and to narrow the pair spacing between each of the three pairs by degree sc is given below:

```
function nlsp=lspnarrow(lsp,sc)
 p=length(lsp);
 wid=diff(lsp);
 %first determine which LSP pairs are narrowest
 n=[pi, pi, pi];
 ni=[0, 0, 0];
 for lp=1:p-1
        if(wid(lp) < n(3))
            if(wid(lp) < n(2))
                if(wid(lp) < n(1))
```

```
                        n=[wid(lp), n(1:2)];
                        ni=[lp, ni(1:2)];
                else
                        n(2:3)=[wid(lp), n(2)];
                        ni(2:3)=[lp, ni(2)];
                end
            else
                    n(3)=wid(lp);
                    ni(3)=lp;
            end
        end
    end
    %narrow the 3 narrow pairs even further
    nlsp=lsp;
    for k=1:3
      nlsp(ni(k))    = lsp(ni(k))    + n(k)*sc;
      nlsp(ni(k)+1)  = lsp(ni(k)+1)  - n(k)*sc;
    end
```

A scaling factor of $sc = 0.2$ or less is usually sufficient to noticeably accentuate spectral peaks, but the code can also be used in reverse, by scaling with $sc = 1.2$ to widen the spectral peaks. Thus, the MATLAB code shown was used to produce the LSPs and spectrum shown in Figure 7.9. Shifting, rather than narrowing, formant-related lines would involve simply replacing the subtraction in the final loop with an addition (although that would neither detect nor correct resonances caused by lines approaching either angular frequency extreme).

Definitely, it should be noted that adjusting lines in this way alters the frequency relationship between any underlying formants, and therefore will tend to degrade the quality of encoded speech. In fact the code shown, when applied to continuous speech recordings, will sometimes result in very unusual sounds once the degree of scaling becomes extreme (for narrowing with $sc \geq 0.5$ or widening with $sc \geq 1.5$), thus it is far better to make only small changes to a speech spectrum wherever possible.

### 7.8.1 Perceptual formant steering

In order to minimise perceived quality degradation, a perceptual basis can be used for line shifting. For example, frequencies can be altered by constant Bark, rather than by a fixed ratio. In this scheme, if $B_k$ is the Bark corresponding to the frequency of a particular line, then that line shifted by degree $\delta$ is:

$$\omega_k' = 600 \sinh\{(B_k + \delta)/6\}. \tag{7.15}$$

To demonstrate this in MATLAB, we would very simply use the bark2f() and f2bark() functions developed in Section 4.3.2 to convert to and from Bark domain, and would apply these to each line in a set of LSPs. So for an original LSP set `lsp`, we can shift by degree $b$ Barks to derive a set of shifted lines `lsp2`. We can shift line $n$ with:

```
lsp2(n)=bark2f(f2bark(lsp(n))+ b);
```

However, it would of course not be necessary to shift every line in the set of LSPs, only the lines directly relating to particular formants would need to be shifted. The hard limit would still need to be applied to prevent LSP values approaching the angular frequency extremes of 0 or $\pi$ and checks made to prevent unintentional resonances caused by moving two lines too close together. As already mentioned, anything other than small changes to a continuous speech spectrum is likely to lead to reduced quality.

## 7.8.2     Processing complexity

In terms of processing complexity, LSP narrowing requires three operations for every line (nine operations for a typical three-formant frame). Shifting using Equation (7.14) requires four operations per line, or 40 operations to shift all lines in a tenth-order analysis frame. LSP shifting using Equation (7.15) requires around six operations per line but, when implemented, will usually necessitate a lookup-table.

Similar formant processing effects can also be produced using adaptive filter techniques. Such a filter requires at least $2NP$ operations per $N$-sample $P$th-order analysis frame.

For tenth-order analysis operating on 240-sample frames, the LSP processes discussed here are between 40 and 400 times more efficient than an adaptive filter. Such figures are only valid where LSP data are available (which is the case for many CELP coders). If LSP data are not available, the overhead of transforming LPC coefficients to and from LSPs would be far greater than any possible efficiency gain.

The methods of LSP adjustment described here have been successfully applied in the intelligibility enhancement of speech [26]. In particular, the use of LSP shifting for altering the balance of spectral power between formant peaks and valleys shows promise. Figure 7.9 illustrates the effect of increasing the separation between the three most closely-spaced line pairs by 20%, and was performed using the `lspnarrow()` MATLAB function on page 190. The figure plots both the resultant spectrum (drawn with a solid line), and the original spectrum (drawn with a dashed line). Traditional bandwidth-altering adaptive filters, such as that of Schaub and Straub [27], perform a similar task, but at a higher computational cost.

Further applications of LSP adjustment may include the quality enhancement of speech/audio and voice masking – described in Section 7.9.

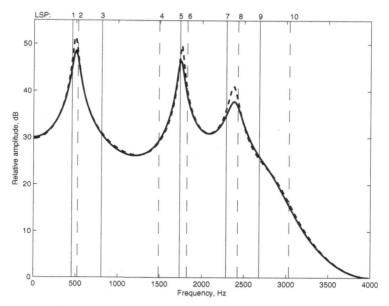

**Figure 7.9** An original spectrum (dashed line) transformed to widen the bandwidth by spreading spectral peaks (solid line) through the use of LSP adjustment.

## 7.9    Voice and pitch changer

If we record some speech at one sample rate and play it back at another, we may notice changes in the perceived frequency of the replayed speech. For example, recording a sentence at 8 kHz and replaying at 12 kHz in MATLAB would cause the output to be obviously different, speeded up in some way. In fact the output would be both higher in frequency, and spoken quicker. However the result probably does not sound like human speech – it has obviously been processed in some way.

The reason is related to the fact that speech is generated by the combination of several physiological processes as discussed in Chapter 3, not all of which are linearly scaleable. Put another way, although the voice of a child may have a pitch rate twice that of a man, the syllabic rate is unlikely to be twice as fast. Furthermore, the frequency location of child formants are unlikely to be twice the frequency of those of a man.

So changing the frequency of a human voice is a non-trivial operation. We cannot simply double everything and expect the output will be convincingly human. Definitely pitch must be scaled to adjust voice 'frequency', but this needs to be adjusted differently from the other components of speech.

Practical methods of vocal frequency translation exist in both the time domain and in the linear-prediction domain. In either case, the important aspect of the technology is to *stretch* the voice of the speaker in some way. Having discussed the importance of pitch in Section 5.3, and the role of the particular shape of the pitch pulse, such stretching or contraction of pitch periods should be accomplished with as little damage to the pitch pulse shape as possible.

There are two ways of achieving this in general. The first, time domain method, is to detect the pitch periods, and scale the waveform shape in a way which is sensitive to the importance of pitch. The second is to completely separate the pitch signal from the speech, and then scale the pitch as required whilst either leaving the remaining speech components untouched, or scaling them in a different way. We will explore two examples: one of a pitch-synchronous time domain scaling, and the other of an LPC-based speech decomposition method. Both methods result in reasonable quality scaled speech that, unless the scaling ratios are very large, can be convincingly natural.

### 7.9.1     PSOLA

The primary traditional method for pitch scaling in audio is known as PSOLA (Pitch Synchronous Overlap and Add). This algorithm lives up to its pitch-synchronous name by first determining a fundamental pitch period. It then segments audio into frames of twice that size, windows them and reassembles the frames using an overlap-add method at a different rate (see Section 2.4 for a discussion on segmentation and overlap) [28].

The different rate of reassembly could either be faster or slower than the original, but as with most such techniques, extreme adjustments can cause significant quality degradation. Figure 7.10 demonstrates the process of speeding up a recording of speech. In this figure, a period of input speech (top waveform) is analysed to determine its fundamental pitch period, $M$. The speech is then segmented into frames of size $2M$ with a 50% overlap, ideally centred on the pitch pulse. Each frame is windowed (see Section 2.4.2), and the resultant audio is 'stitched together' at a different rate. In this case, the pitch pulses are more frequent, thus increasing the pitch rate of the resulting audio.

MATLAB code to demonstrate the effectiveness of the PSOLA algorithm is provided below. This relies upon the function `ltp()` for the pitch extraction method, which we developed in Section 5.3.2.1 to perform long-term prediction (LTP).

Within the code, a Hamming window is applied to frames of size $2M$. The array indexing, using variables `fr1`, `to1`, `fr2` and `to2` to denote the start and end indices of each array, is the heart of the method. This can be applied to a short recording of a couple of words of speech. In this case, scaling of 0.7 will very clearly speed up the speech, and a scaling of perhaps 1.4 will slow down the speech. Note that the intelligibility of the speech remains clear, although the characteristics of the speaker's voice will change.

```
%Determine the pitch with a 1-tap LTP
[B, M] = ltp(sp);
%Scaling ratio
sc=0.35;
```

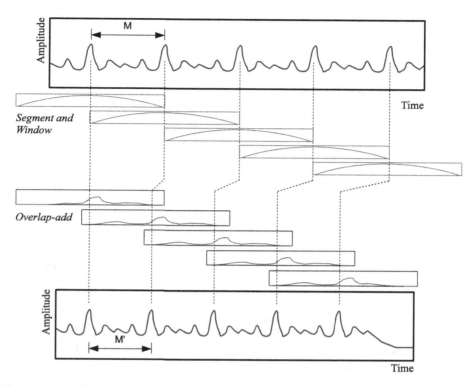

**Figure 7.10** An illustration of the PSOLA algorithm, analysing the pitch period in the top waveform, segmenting it into double sized frames, one per pitch period which are then windowed and reassembled by overlap-adding with reduced spacing to create a faster pitch signal result in the lower waveform.

```
M2=round(M*sc);
out=zeros(N*M2+M,1);
win=hamming(1, 2*M);
%Segment the recording into N frames
N=floor(length(sp)/M);
%Window each and reconstruct
for n=1:N-1
    %Indexing is all important
    fr1=1+(n-1)*M;
    to1=n*M+M;
    seg=sp(fr1:to1).*win;
    fr2=1+(n-1)*M2-M;
    to2=(n-1)*M2+M;
    fr2b=max([1,fr2]);          %Avoid negative indexing
    out(fr2b:to2)=out(fr2b:to2)+seg(1+fr2b-fr2:2*M);
end
```

Most probably, speech scaled by the PSOLA algorithm above, will still sound reasonably true to human speech (in contrast to a straightforward adjustment of sample rate). PSOLA is also reported to work very well with music.

## 7.9.2    LSP-based method

For adjustment of speech, with potentially better performance than PSOLA, the speech signal can be decomposed using a CELP-style analysis-by-synthesis system, into pitch and vocal tract components. The pitch can then be scaled linearly as required, and the vocal tract resonances can also be tugged upward or downward (by smaller amounts) to further scale the speech.

In the CELP vocoder, these alterations can be performed between the encode and the decode process, on the encoded speech parameters themselves. LSP parameter changes (on formant-describing line pairs) are used to tug formants either higher or lower in frequency. Scaling the pitch delay parameter in a one-tap LTP (see Section 5.3.2.1) similarly adjusts the pitch period. With these changes it is possible to shift vocal frequencies, either to change the pitch of a speaker's voice, or to scramble their voice in some way.

A block diagram of a CELP codec modified to perform pitch changing and voice scrambling is shown in Figure 7.11. It can be seen that the encoder and the decoder themselves are identical to the standard CELP coder of Chapter 5, although in practice it would not be necessary to highly quantise the CELP analysis parameters as we would do in a speech compression system.

So if these vocal parameters are not severely quantised, then the targeted adjustments made to the LSP and LTP parameters would cause changes to the processed speech, and these may well be the only perceptible changes made to the processed speech.

These changes are namely the scaling of the LTP delay parameter, and the shifting of formant-describing LSP pairs. For both, a blanket shift throughout a speech recording would probably work, but a more intelligent system which shifted based upon an analysis of the underlying speech, could provide better performance. This could, for example, scale the pitch and shift the formants of voiced speech, but leave the pitch (if any) and formants of unvoiced speech untouched.

Although this technique should properly be inserted into a continuous CELP-like analysis-by-synthesis structure, something of the potential can be demonstrated using the MATLAB code below. In this case, ideally a very short single voiced phoneme such as /a/ needs to be recorded into array sp. We will then perform a single-tap LTP pitch extraction, and an LPC analysis on the voice minus pitch before we scale the LSPs, scale the pitch delay and recreate the speech:

```
%Determine the pitch with a 1-tap LTP
[Beta, tapA] = ltp(sp);
```

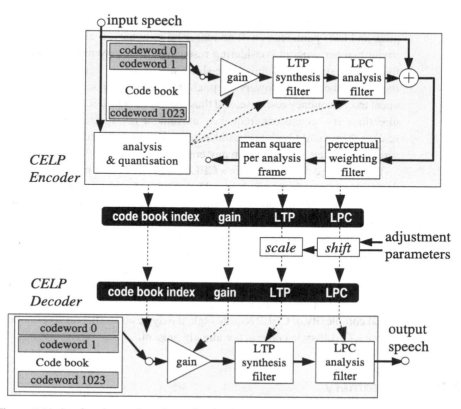

**Figure 7.11** A voice changer based on adjusting intermediate speech representation parameters within a CELP codec, between the encoder and the decoder parts.

```
%Extract the pitch from the speech
vocA=filter([1,zeros(1,tapA-1),-Beta], 1, sp);
%Perform LPC analysis
aA=lpc(sp.*hamming(length(vocA)), 10);
lung=filter(aA, 1, vocA);
lspA=lpc_lsp(aA);
%-- transform --
%lspB=lspA*1.1;
%tapB=round(tapA/1.1);
lspB=lspA/1.1;
tapB=round(tapA*1.1);
%------------------
aB=lsp_lpc(lspB);
%Recreate the speech
vocB=filter(1, aB, lung);
spB=filter([1,zeros(1,tapB-1),Beta], 1, vocB);
```

Remember, this is a simple test only, it will not work over longer recordings because the pitch and LPC parameters will no longer be stationary. It also scales the LSPs in a very trivial manner without considering resonances that may result, or shifts toward either frequency extreme, and thus large shifts will result in significant quality degradation. Increasing the LTP tap lowers the pitch frequency: reducing the LSP values lowers the vocal tract frequency response, and thus the LTP and LSP scaling needs to be in opposite directions if a consistent frequency movement is expected (i.e. for voice frequency scaling they should both move either upward or downward in frequency, but for voice scrambling, they could move in any combination).

Tests of these methods within a CELP coder reveal that pitch scaling and LSP adjustments should remain relatively small but result in fairly natural sounding speech. In very general terms, this method can change a male-sounding voice into a female-sounding one and vice versa. In contrast to the MATLAB code above, the tested system identified the LSP line pairs related to the primary formants (as in `lspnarrow()` on page 190), and shifted these without changing the other lines, wherever possible.

It should be noted at this point that the voice changing method described really does depend upon the presence of a CELP coder. In fact it was designed for use within a telecommunications application as an extension to an existing codec. Since the computational complexity of CELP is quite high, it may be an inefficient solution for applications where a CELP-style coder is not already present.

## 7.10 Summary

This chapter has presented a collection of several methods and techniques that have, in the most part, built upon the speech and audio analysis foundations laid in previous chapters. A complete and workable psychoacoustic model was developed and perceptual weighting discussed (also implemented later using LSP adjustment) along with several discussions relating to the perception of speech and sound. Recent advances in speech analysis and recognition were outlined, along with speech synthesis. Finally, the interesting application of voice masking or pitch changing was discussed, along with two alternative MATLAB implementations of such a system.

## Bibliography

- *Psychoacoustics: Facts and Models*
  Eds. H. Fastl and E. Zwicker (Springer, 3rd edition 2006)

- *An Introduction to the Psychology of Hearing*
  B. C. J. Moore (Academic Press, 4th edition 1997)

- *Acoustics and Psychoacoustics*
  D. Howard and J. Angus (Focal Press, 3rd edition 2006)

- *Hearing (Handbook of Perception and Cognition)*
  B. C. J. Moore (Academic Press, 2nd edition 1995)

- *Speech Enhancement*
  Ed. J. S. Lim (Prentice-Hall, 1983)

- *Music, Cognition and Computerized Sound: An Introduction to Psychoacoustics*
  P. R. Cook (MIT Press, 2001)

- *Speech Communications, Human and Machine*
  D. O'Shaughnessy (Addison-Wesley, 1987)
  A rather expensive book, but one with over 500 pages describing the speech communications field, from basic topics extending to more state-of-the-art coverage of speech enhancement, speech recognition and even a final chapter dedicated to speaker recognition.

- *Survey of the State of the Art in Human Language Technology*
  Eds. R. Cole, J. Mariani, H. Uszkoreit, G. Batista Varile, A. Zaenen, A. Zampolli and V. Zue (Cambridge University Press and Giardini, 1997)
  Also available online from www.dfki.de/ hansu/HLT-Survey.pdf.
  This is a published book, also available online as the result of a joint project between the European Commission and the National Science Foundation of the USA. As the name implies, it describes the state of the art in the human language fields, including speech recognition, speaker recognition and so on.

# References

[1]   C. M. M. Tio, I. V. McLoughlin, and R. W. Adi. Perceptual audio data concealment and water-marking scheme using direct frequency domain substitution. *IEE Proc. Vision, Image and Signal Proc.*, **149**(6): 335–340, 2002.

[2]   I. V. McLoughlin and R. J. Chance. Method and apparatus for speech enhancement in a speech communications system. *PCT International Patent* (PCT/GB98/01936), July 1998.

[3]   B. C. J. Moore. *An Introduction to the Psychology of Hearing.* Academic Press, 1992.

[4]   H. Hermansky. Perceptual linear predictive (PLP) analysis of speech. *J. Acoustical Soc. America*, **87**(4): 1738–1752, 1990.

[5]   Y. M. Cheng and D. O'Shaughnessy. Speech enhancement based conceptually on auditory evidence. *Proc. Int. Conf. on Acoustics, Speech and Signal Processing*, Vol. 2 pages 961–963, 1991.

[6]   Y. M. Cheng and D. O'Śhaughnessy. Speech enhancement based conceptually on auditory evidence. *IEEE Trans. Signal Proc.*, **39**(9): 1943–1954, 1991.

[7]   N. Jayant, J. Johnston, and R. Safranek. Signal compression based on models of human perception. *Proc. IEEE*, **81**(10): 1383–1421, 1993.

[8]   N. Virag. Speech enhancement based on masking properties of the auditory system. *Proc. Int. Conf. on Acoustics, Speech and Signal Processing*, Vol. 1 pages 796–799, 1995.

[9]   D. Sen and W. H. Holmes. Perceptual enhancement of CELP speech coders. In *Proc. Int. Conf. on Acoustics, Speech and Signal Processing*, pages 105–108, 1993.

[10]  A. P. A. Broeders. Forensic speech and audio analysis, forensic linguistics 1998 to 2001 – A review. In *Proceedings of the 13th INTERPOL Forensic Science Symposium*, pages 51–84, Lyon, France, October 2001.

[11]  A. P. A. Broeders. Forensic speech and audio analysis, forensic linguistics 2001 to 2004 – A review. In *Proceedings of the 14th INTERPOL Forensic Science Symposium*, pages 171–188, Lyon, France, 2004.

[12]  S. Furui. Recent advances in speaker recognition. *Pattern Recog. Lett.*, **18**: 859–872, 1997.

[13]  S. Furui. Speaker-dependent-feature extraction, recognition and processing techniques. *Speech Commun.*, **10**: 505–520, 1991.

[14]  G. Doddington, W. Liggett, A. Martin, M. Przybocki, and D. A. Reynolds. Sheep, goats, lambs and wolves: A statistical analysis of speaker performance in the NIST 1998 speaker recognition evaluation. In *Fifth International Conference on Spoken Language Processing*, volume 0608, November 1998.

[15]  M. A. Zissman and K. M. Berkling. Automatic language identification. *Speech Commun.*, **35**: 115–124, 2001.

[16]  R. Cole, J. Mariani, H. Uszkoreit, G. B. Varile, A. Zaenen, A. Zampolli, and V. Zue, editors. *Survey of the State of the Art in Human Language Technology.* Cambridge University Press, 2007.

[17]  G. A. Miller, G. A. Heise, and W. Lichten. The intelligibility of speech as a function of the context of the test materials. *Experim. Psychol.*, **41**: 329–335, 1951.

[18]  C. A. Kamm, K. M. Yang, C. R. Shamieh, and S. Singhal. Speech recognition issues for directory assistance applications. In *Proceedings of the 2nd IEEE Workshop on Interactive Voice Technology for Telecommunications Applications IVTTA94*, pages 15–19, Kyoto, Japan, September 1994.

[19]  W. Walker, P. Lamere, P. Kwok, B. Raj, R. Singh, E. Gouvea, P. Wolf, and J. Woelfel. Sphinx-4: A flexible open source framework for speech recognition, 2004. URL cmus-phinx.sourceforge.net/sphinx4/doc/Sphinx4Whitepaper.pdf.

[20]  M. Edgington. Investigating the limitations of concatenative synthesis. In *EUROSPEECH-1997*, pages 593–596, Rhodes, Greece, September 1997.

[21]  T. Dutoit. High quality text-to-speech synthesis: An overview. *Electrical & Electronics Engng, Australia: Special Issue on Speech Recognition and Synthesis*, **17**(1): 25–36, March 1997.

[22]  The University of Edinburgh Centre for Speech Technology Research. The Festival speech synthesis system, 2004. URL http://www.cstr.ed.ac.uk/projects/festival/.

[23]  P. Taylor, A. Black, and R. Caley. The architecture of the Festival speech synthesis system. In *Third International Workshop on Speech Synthesis*, Sydney, Australia, November 1998.

[24]  K. K. Paliwal. On the use of line spectral frequency parameters for speech recognition. *Digital Signal Proc.*, **2**: 80–87, 1992.

[25]  I. V. McLoughlin and R. J. Chance. LSP-based speech modification for intelligibility enhancement. In *13th International Conference on DSP*, Santorini, Greece, July 1997.

[26]  I. V. McLoughlin and R. J. Chance. LSP analysis and processing for speech coders. *IEE Electron. Lett.*, **33**(99): 743–744, 1997.

[27]  A. Schaub and P. Straub. Spectral sharpening for speech enhancement/noise reduction. In *Proc. Int. Conf. on Acoustics, Speech and Signal Processing*, pages 993–996, 1991.

[28]  H. Valbret, E. Moulines, and J. P. Tubach. Voice transformation using PSOLA technique. In *IEEE International Conference on Acoustics, Speech and Signal Proc.*, pages 145–148, San Francisco, USA, March 1992.

# Index